# From Quarks to Black Holes

## Interviewing the Universe

# From Quarks to Black Holes

## Interviewing the Universe

**Richard T. Hammond**
*North Dakota State University*

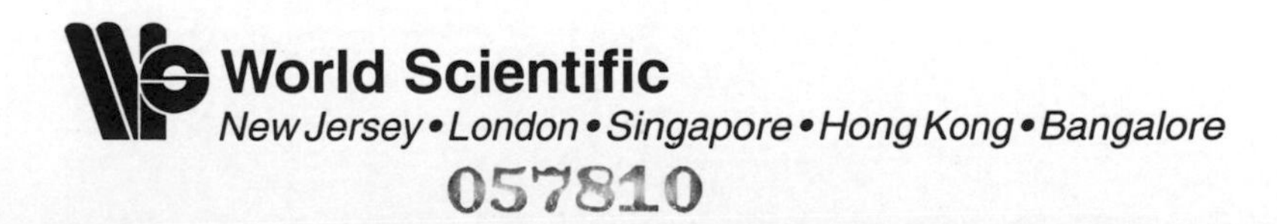

*Published by*

World Scientific Publishing Co. Pte. Ltd.

P O Box 128, Farrer Road, Singapore 912805

*USA office:* Suite 1B, 1060 Main Street, River Edge, NJ 07661

*UK office:* 57 Shelton Street, Covent Garden, London WC2H 9HE

**British Library Cataloguing-in-Publication Data**
A catalogue record for this book is available from the British Library.

**FROM QUARKS TO BLACK HOLES: INTERVIEWING THE UNIVERSE**



ISBN 981-02-4625-0

Printed in Singapore by Mainland Press

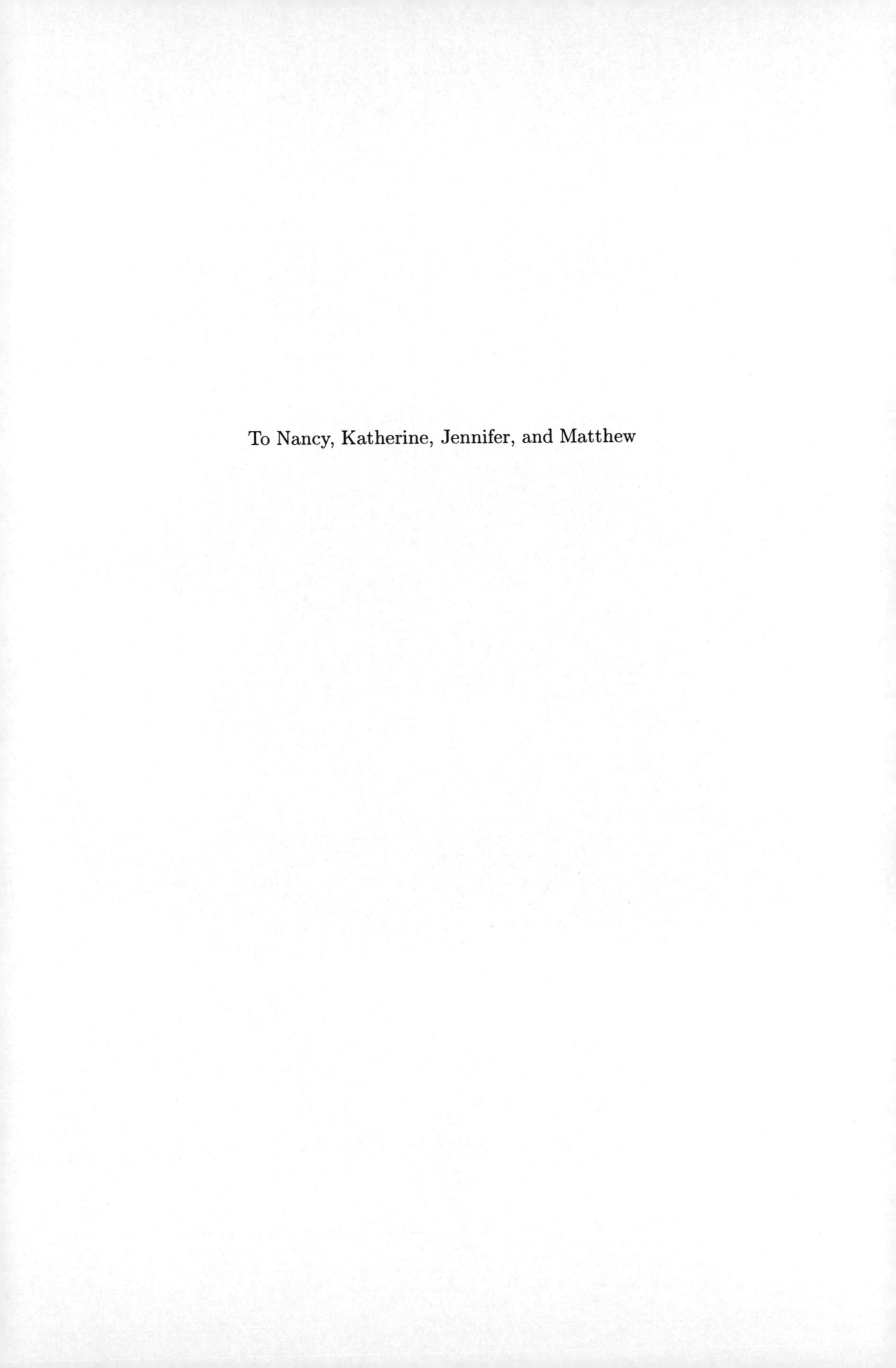

To Nancy, Katherine, Jennifer, and Matthew

# Preface

I must admit that the idea of interviewing natural objects was not my own, and in fact would have thought it to be quite impossible, had it not been for a conversation I struck up with a carbon atom. As she told me of her spectacular creation in a remote part of the Universe, her incredible release in a supernova explosion, and her experiences on Earth, I could barely contain my excitement. It took me a while to reach a quasi equilibrium, but when I did I found myself furiously jotting down notes. It was such a wonderful experience I could not suppress the desire to recapture the moment, and to my glee, found an electron, who was much younger, equally willing to share his experiences about his creation in the upper atmosphere, his adventures in household appliances, and his hair-raising brush with annihilation in an accelerator. I jumped at the chance when Jupiter volunteered for an interview, although was taken unawares at his unhappy nature, still brooding at his inability to have been a star.

Buoyed with my initial successes, I ventured out a little further, and struck up a conversation with a black hole. This was extremely difficult, and most of my notes are gibberish, as I mistakenly tried to write down many of the equations she put on the blackboard. Unfortunately, I saw many objects falling in, and fearing for my own safety, was forced to bring the interview to a close. The interview with the uranium atom was equally surprising, but for different reasons. It was deeply concerned about being used in weapons of mass destruction, and I found myself on the defensive, surprised that an atom's concern that, at times, outweighed those of humans.

By now the word was spreading, and I had many visitors knocking at my door. In order to accommodate the growing queue, I decided to interview two at one time. This was a mistake, and although the fermion was well behaved, the boson was too self-centered and very rude. Nevertheless, they provided a good discussion on the roles they play in nature.

The star I interviewed is, as you might suspect, our sun, but as you might not suspect, after emphasizing that it took the entire universe billions of years to bring together the elements we find on Earth, chides us for wasting this precious gift.

The hydrogen atom is also somewhat surprising, and discusses the philosophical implications of quantum mechanics, as well as its more concrete predictions. The quark, however, takes us even deeper and tackles the issue of beauty in physics.

These objects and all the others I interview each have a unique perspective of the world in which we reside, and I would like to take this opportunity to thank them all for exposing their inner thoughts.

# Contents

List of Interviews

**Interview with a star** 34

We find that this star happens to be our sun, who tells of his birth, life, and looks into his death. The Sun describes some of his properties, such as size and mass, sunspots, and brightness, and describes the war raging within his core. Looking forward, the Sun discusses his ultimate fate, describing red giants and white dwarfs.

**Interview with a Wimp** 44

This weakly interacting massive particle, which happens to be a neutralino and is not exactly happy with its wimpy moniker, explains what weakly interacting particles are, discusses the fundamental symmetries of elementary particles, and explains supersymmetry.

**Interview with a comet** 51

Unlike previous "interviewees," we find the comet to have a keen interest in the history of scientific development during the last two millennia. The comet relates scientific achievements on Earth to its orbital place in the heavens, and we see that the comet shares the concerns of the uranium atom concerning the human's proclivity toward self-destruction.

**Interview with a spiral galaxy** 58

A spiral galaxy discuss its structure, including the great spiral arms, and reveals one of our greatest mysteries. It explains that in order to account for the motion of the stars and gas in its outer regions, there must be huge quantities of invisible matter filling the galaxy. Despite prodding, the galaxy will not tell us what this dark matter is, leaving us with only a few tantalizing hints.

**Interview with a neutrino** 67

The neutrino shares her excitement with being discovered, and, after a discussion about energy and matter, explains why it takes 100,000 gallons of cleaning fluid to find her. The neutrino also explains the solar neutrino puzzle, one of the leading unsolved mysteries in physics and astronomy.

**Interview with a hydrogen atom** 77

After describing its discovery, the hydrogen atom brags about the radiation it emits, and introduces us to quantum mechanics, the laws of physics that rule on the atomic scale. It discusses the concept of discrete versus continuous, and destroys our notions of a deterministic world.

## Interview with a neutron 90

The neutron, worried about his own demise, stops by for a short interview. The neutron discusses his makeup, desire to join a nucleus, and gets into a discussion about particles and waves.

## Interview with a quark 98

The quark explains why there was initial skepticism to its existence, and discusses why it appears impossible to ever observe a lone quark. The quark talks briefly about flavor and color, enters into a discussion about faith and beauty, and the fundamental role they play in understanding nature.

## Interview with a tachyon 107

For the first time we see some skepticism of our friend, the interviewer. We find that if tachyons, particles that travel faster than the speed of light, exist, then the cherished notion of causality is brought into question.

## Interview with a quasar 111

Quasars puzzled astronomers for many years, and in this interview we see they afford us a view into the past. These objects are the most energetic in the Universe, and have a unique engine that generates their enormous power.

## Interview with antimatter 116

An antielectron stops by to explain the properties of antimatter, and mentions that antihydrogen has already been produced in the lab. The discussion turns to how antimatter may be used in rocket propulsion, and a brief discussion on negative mass follows.

## Interview with an iron atom 122

The iron atoms explains its supernova origins, how it came to Earth and surfaced from erosion. It describes its participation in early wrought iron processing, its role in the human body, and its inclusion in modern high grade steel.

## Interview with a muon 131

The muon, which compares itself to an overweight electron, has little time to spare. It gives a brief description of how it was found, explains a little more about exchange particles and the origin of the nuclear force, and goes on to discuss relativistic length contraction.

## Interview with a neutron star 137

The neutron star explains a supernova explosion and its personal origins. It also discusses pulsars, x-ray bursters and gamma ray bursters.

## Interview with a string 145

A string compares itself to the more common view of elementary particles as points, and explains how we can exist in more dimensions than we see. The string also compares classical and quantum theories, and claims to be the only possibility to describe a quantum theory of gravity. Finally, the discussion comes around to beauty, and we hear the string's honest comparison of his world and the standard view.

## Interview with vacuum 153

We find the vacuum to be a very active arena, and as it describes its unique properties we begin to realize how important vacuum is. Vacuum explains a little more about Einstein and his theory of gravity and the expansion of the universe, and both chides and praises our research into natural law. Vacuum closes with a personal statement on its perspective of the Universe.

## Interview with a muon

The muon, which compares itself to an overweight electron, has little time to spare. It gives a brief description of how it was found, explains a little more about exchange particles and the origin of the nuclear force, and goes on to discuss relativistic length contraction.

## Interview with a neutron star

[illegible]

## Interview with a string

[illegible]

## Interview with vacuum

[illegible]

## 0.1 Interview with a carbon atom

*I'd like to thank you for this interview. It's a first for me, and I'm not sure I will ask you all the correct questions, so please feel free to improvise. I'll begin by asking you where you are from.*

You might be surprised that I remember my birth so well, particularly since it was so long ago and far away. Nevertheless, after waiting what seemed like an eternity, my turn came. My parents, who numbered more than two, finally felt enough heat to combine and, through nature's reincarnation, I was born — a carbon atom from three helium atoms.

*Do you mean that you are the product of fusion?*

Yes, quite so, although excitement with my new life began to fade as quickly as my star cooled, and I came to the realization that I would be forever trapped inside a giant, inert star made of nothing but carbon copies of me. However, my star was inviting, through its extensive gravitational influence, huge numbers of sister atoms from our companion star, and there were rumors that our comfortable society would collapse.

*So, after there was no more helium to make carbon, fusion stopped, and then you drew in atoms from an orbiting star?*

Yes, I guess I was lucky. The rumors proved to be made of sterner stuff than we, and at one fateful moment, long before you began to measure time, the gravitational field became so strong that none of us could bear it. We collapsed down to a size that, to this day, makes me shudder to think about, and before any of us knew what happened, we blew apart in the biggest and most spectacular explosion in the

Universe. I was fascinated by the process, not only because for the first time was I free, until then, the entire Universe housed nothing but hydrogen atoms (my grandparents), helium, and carbon; but now, as I was zooming along traveling at nearly light speed, I saw all kinds of heavy and strange elements. I learned fast that these fearful newcomers could swallow me whole without a second thought, and tried to steer clear.

*Are you describing a supernova and its aftermath?*

Yes. Afterwards, thousands of years shot by like a day, and millions turned to billions as once again I was caught in a tedious monotony. Far from home, and in stark contrast to my earlier heated environment, I found myself trapped in a cold, gloomy expanse with my nearest neighbors, hydrogen atoms, much too far away to communicate.

*Sounds dreary. How did you buoy your spirits?*

With the passing of a few more million years, my despair lifted as I noticed a few neighbors coming my way, and then a few more, and soon we were knocking around like old friends. That was when the rumor mill fired up, spreading news of that spectacular event — collapse.

*But this is a different kind of collapse than the one you mentioned before. Now you are saying there were rumors of star formation?*

They proved accurate, and in another few million years we were whirling around, inching closer to what we all knew would be a star. Most of my new hydrogen friends made it in, but I got stuck in orbit, and although I witnessed the birth of the star from afar, I felt left out, too old and heavy to be of service.

*What do you mean by too heavy?*

In a new star hydrogen undergoes fusion to make helium, and much later helium makes carbon. I am about twelve times the mass of hydrogen, and more of an end product than real fuel of the heavens.

*What did you do?*

Fear broke my shell of self-pity like a nutcracker. Although I was in orbit, I was not alone. Many heavy elements and even those awful molecules were gathering around me, trying to steal my electrons while choking out the new sunlight. In no time I was buried deep within a solid ball of iron and minerals. I could not begin to measure time in that terrible blackness, pushed and shoved from all sides with nowhere to go but eternity.

*Do you mean you became part of the Earth?*

Yes. Millions of years piled up like old tires in the junk yard, but finally, out of nowhere came an enormous seismic shock, so powerful that it ripped open the terrible ball and I found myself once again in space, orbiting not one but two bodies. Soon my orbit degraded and I found myself on the larger body, and the other, which impacted us, fell into orbit. Again I found myself in strange society with many visitors from space. They would smash into my planet with alarming rapidity, and being made largely of water, their remains soon formed giant oceans and I feared any millennium I would drown.

*I believe that you are describing the birth of our Moon, and how comets bombarded the planet, but I would hardly think you could drown.*

Just a little poetic license. Then came the strangest thing of all — I took part in the most exciting project that ever occurred in the Universe. The fusion, supernova, solar system formation, birth of your Moon, and comet bombardment all paled in comparison to this unique experience. First, to my initial horror, I became part of a molecule, but then we, and many others, formed a most intricate, and I would say, unnatural shape, with long tunnels bringing water upwards, against gravity, and green ethereal pages flapping in the breeze and absorbing sunlight, the same light that my ancestry once produced. I loved the irony, and was fascinated by the fact this structure could react and respond to that very light.

*It sounds as though you became part of a plant of some kind.*

A tree. However, this wondrous time was short-lived. Soon, no water could be extracted from the soil beneath, and the sunlight fell on an unresponsive structure. A great paralyzing sadness, like no other I ever felt, covered me up as completely as the surrounding Earth. Deeper I went, and time, as powerful as any force in the Universe, stripped me of my neighbors, and although I would pick up some hydrogen (four of them and I would often travel together), I found myself barely moving through a thick black sludge for, again, more years than I can remember.

The black gloom was broken by a sudden change in pressure. We all felt it at the same time, and went rushing up, blindly, and I found myself on the surface again, this time trapped in a strange, geometrical shape, something I thought was beyond nature's talent. But that was nothing compared to what I was about to see and experience. Inexplicable, terrifying, and unpronounceable structures of all sizes were everywhere, and I was refined, reused, abandoned, salvaged, and, in general, forced into a vast multitude of objects and shapes I can hardly begin to explain. Things were happening so fast I wished I could have kept a diary, but being a carbon atom, well, it's difficult.

*Wait a minute. You died, I mean your tree died, decomposed to oil, and then what? Recently you came back to the surface from an oil well?*

Mmm-mm. I was totally absorbed in trying to understand humans, what you are and where you come from, and for a short while I was even part of some bizarre scheme in which I found myself wrapped around a woman's leg — from her feet right up to her waist. I saw some strange things during that period, but I don't think I should relay them. In fact, most of the applications I participated in are beyond my comprehension. Although the principle of fusion is clear to me, most of my recent history is not.

I'm not complaining, I love being part of such a dynamic, fast-paced life, but a shadow of concern passed over me as I found myself, just the other day, immersed in black fluid, again. This time, however, we were quickly shipped off to a publication plant and before I could dot an *i*, here I am, in the middle of the period at end of this sentence. Nice to meet you.

*What an exciting life, what do you propose to do in retirement?*

I cannot begin to imagine what the next ten billion years will have in store for me. If I could bare my soul I would tell you that, deep down, I would give up my inner electron to be part of a human, short-lived as you are. Perhaps I am just reaching for the stars.

*Perhaps not.*

### 0.2 Interview with an electron

*Thank you for granting me this interview. I know you want to be on your way, so I'll...*

No, I would be happy to stay right here and rest a while. I'm always on the move — a rest would do me good.

*Well, make yourself at home. To begin, would you tell us your age and where you are from?*

Born right here in the U.S. of A., 'bout fifty years ago.

*How did this come about?*

Pretty common, I must admit. An alpha particle from the Sun...

*An alpha particle?*

The nucleus of a helium atom, you know, two protons and two neutrons. Well, it was emitted from the Sun along with countless others. It smashed into a nitrogen atom about 250 kilometers up. Going at over 100 million meters per second, I can tell you, there was nothing left of either atom. Protons, neutrons, and many other particles were scattered across the sky like stars on a clear night. There was still plenty of energy left over, so I was created from that energy along with my alter ego, a positron. Unfortunately for him, he smacked into another electron and was annihilated. I was captured about 100 kilometers up by an oxygen atom, and finally worked my way down to the surface.

*Then what?*

Rust.

*Rust?*

You know, oxidation. It was an old abandoned oil rig in Texas. My oxygen molecule...

*I thought you said oxygen atom?*

Originally, yes, but as I descended down into the lower atmosphere, and became shielded from the ultraviolet rays of the Sun, we bonded with another oxygen atom and became a molecule. I didn't like that at all, I was being shared by two different atoms, first belonging to one and then the other, back and forth, back and forth, never really knowing to which atom I belonged. After a while, though, I grew accustomed to the new life, and even enjoyed some of the added excitement.

*I see, now, about your oxygen molecule?*

My oxygen molecule — I don't know what it was thinking — got too close to an iron atom; it reached out and grabbed us like a frog snatching a fly. I'll bet that poor old oxygen is still there.

*How did you get away, what changed?*

My whole life-style changed radically. With oxygen, I had a home, good neighbors, and although we would get excited from time to time, we had a stable life. In iron, I was continually jostled from one atom to the next, no atom could really give me a permanent home, and I became a vagrant. The slightest electric field would send me zooming, knocking into my brothers like morning commuters at Grand Central station.

*It sounds like you are describing how electricity flows, but how did you escape the oil rig?*

It didn't take long. A thunderstorm passed to the north of us, and there was an accumulation of positive charge. Now, you probably know, if there is one thing we electrons cannot resist, it's positive charge — we fly to it like bees go to flowers. Next thing I knew, I was in the ground buzzing north, along with about a zillion brothers.

*A zillion?*

Its hard to count, and I usually don't bother, but certainly we numbered well over $10^{25}$. It was a very treacherous trip, I might add.

*How so?*

A lot of us didn't make it — trapped by atoms and molecules. I don't want to dwell on the negatives, so I should tell you about my household adventures.

*Please do.*

Traveling north, as I said, I became part of the local power grid.

*Wait a minute, I thought commercial electricity is sent through wires.*

Yes, but up to 30% is actually carried by ground currents — don't blame me, you designed it. In fact, as I was traveling near a dwelling, before I knew what hit me, I was pulled into the house and went through every electrical device in the house, from vacuum suction devices to color TV.

*How did you like being in TV?*

Disappointing. I was hoping to get the free ride, but I was routed back into the household wiring.

*Free ride?*

That's what we call it. In the picture tube you get accelerated through a few thousand volts and then ejected into vacuum and travel clear to the screen. I was hoping not only to get the free ride, but to shoot out and grab onto an oxygen molecule, or even a nitrogen molecule. But I went back into the wiring. It was not pleasant, either. You use alternating current and we all jiggle back and forth like jello in an earthquake.

*How do make any progress at all?*

Sheer willpower.

*Willpower?*

Just a joke. Now and again there are net potential differences, which means that on average, for a while, we go more in one direction than the other. One day I went from a basement sump pump to the kitchen toaster, which was my last appliance.

*What happened?*

Well, I liked the toaster, but its owner jammed a bagel in there and a crumb got very close to the heating element. The element got red hot, boiled me off, and before I could get my bearings, I was part of the bagel. Things got a little crazy after that. He ate me, of course, and I got banished to Siberia.

*Siberia?*

His hair, that's what we called it. It was pretty barren up there. You know what I mean.

*Yes, I do.*

Well, his wife cut his hair, what little he had, on their back deck. I blew along the ground, looking for an opportunity to get back into the ground current, but a bird got me and we became part of a nest. Quite disgusting. She finally abandoned it, I got washed away, and before winter hit I was in the Gulf of Mexico riding a current that would take me to the shores of France. I could write a book about what happened to me in the ocean, but I soon ended up in Switzerland and came face to face to with our ultimate fear — annihilation. I know human's often brag about the great facilities at CERN, but to us, it is our worst nightmare.

*I know that CERN is the European Organization for Nuclear Research, but what were you doing there?*

It was certainly not my choosing. Don't forget, we must obey orders, and when electric fields say march, we march. Yes, we muster some resistance, and we don't always let you know just where we are or precisely what we are doing, but on average, we are helpless against overpowering fields that drive us. The project was called ALEPH, which, among ourselves, was called the firing squad. First they would get us to go around this great circle, 27 kilometers in circumference, accelerating all the time until we nearly reached the speed of light. At that speed we measured the entire 27-kilometer length to be about half a foot.

*You are describing relativistic length contraction?*

Yes, but the main point is this: going in the opposite direction were our alter egos, positrons. You remember I described my birth, well, we die in just the opposite way. If I get too near a positron, it's curtains. We both get obliterated, leaving behind, at that energy, all kinds of other particles. By the way, positron annihilation is one of the few ways we can be destroyed: Left to ourselves, we'd live forever.

*I would think that a head-on collision is very rare, though.*

It is, and that's the nightmare. They made us go around and around, over and over, until nearly all of us were destroyed. It was a bloodbath.

*But, if I remember correctly, didn't these experiments...*
Experiments!

*Sorry, but weren't the Z and W particles created, confirming the theory relating to the weak nuclear forces?*
Yes, this is true, and in some circles we have come to view our lost comrades as patriots and martyrs.

*I must admit, I've never thought about it from your reference frame. Obviously you escaped, how?*
One of the superconducting magnets got too warm, weakening its magnetic field, and I strayed into the collider wall, worked my way along an aluminum shielding cable, was recycled in a restoration of the facility, and found myself in the wing of an airplane. I traveled around quite a bit since then — I could write another book — but finally agreed to your interview.

*Well, thank you for relating your experiences to us, I would like to wish you luck in the future.*
Thanks, I'm looking forward to reading your book.

## 0.3 Interview with Jupiter

*I usually begin these interviews by asking a question about birth. Would you comment on yours?*

Well, that was about five billion years ago, and I've forgotten some of the details. The general rumor is that a great cloud of hydrogen was collapsing and shrinking fast. The smaller it got, the faster it rotated until finally a large chunk of it spun out of control and was cast out into a distant orbit around what finally became your Sun. I am the outcast that never made it in.

*You should cheer up, people have marveled at you for centuries; you were named after the most powerful of the ancient gods.*

I am nothing but a failed star.

*A failed star?*

Yes. I am mostly hydrogen, just like your Sun. If I were bigger, fusion would have begun and I'd be a star. Instead I am a dud, a failure.

*Why would you be star if you were bigger?*

I am very hot in the center, but not hot enough. This is heat left over from my birth when all the hydrogen was crashing together. But in order to make fusion, where protons and neutrons bind together to make helium, I would have to provide them with better introductions.

*Better introductions?*

I mean I would have to force them together, get them closer. To do that I would have to apply more pressure at my center, and to do that I would need quite a bit more mass. But I am deficient in this, and doomed to shine only in another's light.

*I wish I could cheer you up a little. I should emphasize that you are revered on Earth.*

Reverence or pity?

*It is not pity, I assure you. But, tell us a little more about yourself. What about your moons?*

They are my pride and joy, I must admit. My one solace in a desolate life. My four largest moons, discovered by Galileo, are like my children. I love my many others as well, but they never achieved the stature of Io, Europa, Ganymede, or Callisto.

*I understand Io has volcanoes. This was a great discovery here on Earth. Since Io is only slightly more massive than our Moon, we thought it would be dead, geologically inactive, I mean.*

I keep Io hot through my tidal force. The heat I create causes the volcanoes.

*So, it's not like on Earth?*

No, not at all. Out in these bleak surroundings we try to make the best of a miserable life.

*Can you explain what you mean by tidal force?*

Let me see. You are aware that your single Moon creates tides on Earth. This is because the side of your tiny planet that is nearer to the Moon experiences a stronger force than the side that is farther away. The net effect is to tend to pull the planet apart.

*And this is the origin of tides on Earth?*

Yes. On Io, my tidal force actual distorts the moon's shape. Take a tea spoon and bend it back and forth a few times and notice how hot it can get. That is what is happening on Io. It is one of our few enjoyments in the barren depths of space I inhabit.

*You also have a ring, like Saturn.*

No, not like Saturn. My ring is so small and thin you didn't even see it until 1979, when the spacecraft Voyager paid me a visit. It's the story of my wretched life, too little too late.

*What about your great red spot? It's as big as Earth.*

Phmmmp. A mere storm, it'll be gone in the blink of an eye.

*Well, you have the most striking colors of any planet, pink hues to darker reds, tawny brown, vivid yellows, all arranged in bands across your surface. How do these come about?*

Warm material rises in zones, which are relatively bright. The gases cool down and sink, forming darker cool belts, back into the interior. These are all adjacent, so you see bands. It is a little bit like your trade winds, but of course, in this sorrowful backwater, no mariners will ever set sail, no...

*Ahemm, excuse me, I must be getting a slight cough, but what causes those wonderful colors?*

Sulfur, mostly. A little goes along way.

*I see. Well, you certainly are magnificent to look at. I wish there was a little something I could do to cheer you up.*

Voyager was nice, and the other spacecraft too. I appreciate the visits, perhaps a little more often?

*I'll see what I can do.*

Thank you.

*Thank you.*

## 0.4 Interview with a black hole

*Please excuse me if I keep my distance.*

I quite understand. In fact, I was going to warn you that if you did venture within my event horizon, you could not escape.

*Perhaps you could talk about this event horizon a little.*

Well, let's start from Einstein's field equations of general relativity, which relate the contracted Riemann tensor to the energy-momentum...

*Ah...excuse me. Would you be willing to discuss this in a less technical way?*

No problem. How's your imagination?

*Okay, I guess.*

Good. Picture a large, black balloon. Picture this black balloon a meter across, now two meters across, so its radius is one meter. Now imagine blowing this balloon up until it has a radius of 10 meters, now 100 meters, now a few kilometers. Are you with me?

*So far.*

Good. Now imagine that the entire mass of the Sun is at the center of the balloon, squeezed into a point. Still with me?

*Yes, I'm trying.*

Now imagine that the black balloon surface will allow anything to pass from the outside to the inside, but nothing from the inside can leave that surface. Not particles, not light, nothing.

*Like a one-way membrane?*

Yes. Now you understand me. The imaginary surface of the balloon is what we call the event horizon. Another view is this; it's the point of no return.

*If I had very powerful rocket engines, though, couldn't I escape then?*

No, sorry. Once you're inside the event horizon, your fate is sealed. No matter what you do you will end up at the center. We call that point a singularity — it's nasty in there, and of course you would not survive.

*If nothing can escape, how were you discovered?*

I have visitors knocking on my door all the time. In fact, there's so many trying to get in there's a traffic jam out there worse than the Long Island Expressway at five in the afternoon. They get pretty hot under the collar, let me tell you. In fact, the surrounding material gets so hot it emits a particular kind of x-ray radiation. That is what you see and that is what you measure.

*So we don't really see you, just the gas outside your event horizon, and that's the only way we know you exist?*

Yes and no. I happened to be part of a binary star system.

*You mean two stars, or a star and yourself, rotating around each other?*

Yes, and most of the material passing across my event horizon comes from that star. The first thing you noticed was my companion star moving in an orbit around something that wasn't there. Of course I was there, but you didn't see me at first. By calculating where a massive object must be to account for that motion, you began to look for the x-rays, observed them, and the rest is history.

*This is a lot to take in. How is it that one of the binary system is a star, and the other, I mean you, is a black hole?*

It's a long story.

*I've got all the time in the world.*

I could say that too, but I'll try not to use it all in my explanation.

*Thanks, I guess I was exaggerating.*

I know, but I wasn't. We formed from a giant hydrogen cloud, similar to the birth of your solar system, as explained by Jupiter, but on a much larger scale. By the way, what's got Jupiter so bummed out? Poor kid.

*I don't know, his lugubrious nature surprised me. But you were saying?*

Oh yes, sorry. As the hydrogen cloud collapsed we formed into two stars rotating around our common center of mass. My companion is about twice the mass of your Sun and I started out as a very happy and bright star more than three times the mass of your Sun. Being bigger I burned faster, hydrogen to helium, and then helium to carbon. I started to cool down as the fusion ceased, but our orbits brought us pretty close, and I started attracting matter off my companion like a dog pulls in fleas. The fatter I got the more I pulled in, and finally the carbon atoms could not hold out against the enormous gravitational pressure. We collapsed like a popped balloon, crunching down to solid neutrons. For a moment I thought we would become a neutron star, supported by neutron degeneracy, but...

*Please, not too technical...*

Sorry, you should interview a neutron star for more details about that, but the gravitational field was too strong for even solid neutrons. So there was an enormous explosion. Much of the matter was squeezed way beyond its natural density, and when it expanded it produced the largest explosion since the Big Bang. Not only did it explode, the enormous energy produced, or brought, all the heavy elements that you find here on Earth.

*Wait a minute, you mean that the iron, the uranium, the gold, the lead, all that, came from this explosion?*

Yes, or from one like it. In fact your carbon atom explained it, but she was a little too poetic for me.

*And after the explosion?*

Well, after the explosion, a little less than half of the mass stayed behind, and kept collapsing. Then, a trapped surface was formed...

*Trapped surface?*

Oh, sorry. Basically, the next thing I knew — I had an event horizon. It took me quite a while to understand what was going on. I could see out, but nothing could see in. Anyway, the matter collapsed down to a singularity, I mean a point, and there you are, I was formed.

*What a birth! But can't anything halt that collapse process?*
No, gravity overpowers all other forces.

*I see. Are most black holes your size?*
Well, there are many like me, but you have been having more luck finding the big ones.

*How big?*
Well, I'm about one solar mass, but lately you have been observing black holes between 10 million and 100 million solar masses. They're actually much safer.

*Safer?*
Yes. You could get very close to the event horizon and be perfectly safe. With me, if you got close to my event horizon the tidal force would pull you apart long before you got through.

*Tidal force?*
The Riemann tensor contracted with the...

*I think you're losing me again. Is that the same tidal force Jupiter exerts on Io.*
Yes, that's right. Only since my gravitational field is so much larger than poor old Jupiter's, I would do much more damage. Here, I'll show you... No no, don't jump back, may I use your black board?

*I'm sorry, but I really don't understand all those equations very well.*
Okay, I'll put it this way. If you fell through my event horizon, the tidal force would be about about two trillion tons, that means there would be a force of about two trillion tons trying to pull you apart.

*Dear me, but its not so bad for a large black hole?*
No, you could fall through unscathed, and in fact, you would have plenty of time to enjoy the ride in.

*I'll bear that in mind. By the way, I have heard of the phrase, naked singularity. Do you know anything about that?*
As I explained, I have a singularity at my center.

*Perhaps you could remind me, one more time, what you mean by singularity.*

Well, look here on the black board, look what happens when the radius goes to zero.

*I see, but in plain words how would you explain this.*

Well, when you have a large amount of matter, in my case about the mass of your Sun, which is $2 \times 10^{30}$ kilograms, all located at a single point, that's a singularity. However, you cannot see that singularity because it is cloaked by my event horizon. If I had no event horizon my singularity would be visible to the outside world, it would be naked.

*So a naked singularity is a singularity without an event horizon?*

Yes.

*Is this possible?*

Not according to Einstein's General Theory of Relativity. It was proven that if a singularity is formed, then there must be an event horizon.

*Then why all the speculation about naked singularities?*

The proof is based Einstein's equations, and certain assumptions about the energy and pressure of matter. If these assumptions are wrong then the proof will not be valid. You must keep an open mind about these things, you know.

*Do people believe in them or not?*

Mostly not, in fact, Roger Penrose thinks they are inevitable and coined the phrase, cosmic censorship, which forbids their existence.

*I was wondering about something else. What happens if two black holes collide?*

You get a bigger black hole.

*Oh, and what happens to you as more and more matter falls into you.*

I get bigger. The radius of my event horizon is directly proportional to mass, here let me show you on the black...

*No thank you, I understand. So you can just keep growing and gobble up an entire galaxy, and then...*

No, once you get far away from my event horizon my gravitational field weakens just like that from any other object. However, some black holes near the centers of galaxies gobble up other stars, and other black holes.

*So, the only thing we can measure from a black hole is the gravitational field?*

No. You can measure my angular momentum and charge also.

*How can we measure your angular momentum?*

It is difficult, I admit, but as we spin, we not only drag the in falling particles around, we drag a little space too.

*Drag space?*

Yes, you call it the dragging of inertial frames, kind of a silly name. But it leads to observable effects.

*And charge?*

Well, most of us are pretty neutral, but we can have a net positive or negative charge. You can measure our electromagnetic fields.

*I'm confused. If nothing can escape from a black hole, how does the electric field get out?*

It's already there when we collapsed. By the way, I never claimed that I could not influence the space around me. My gravitational field as well as an electromagnetic field were present before I ever had an event horizon. Just because the event horizon formed does not mean that it cut me off completely from the rest of the Universe. None of my charge can escape, none of my matter can escape, but my fields exist throughout space much like they did before the event horizon was formed.

*I see. There is one more topic I would like to broach.*

Broach away.

*I have heard the phrase worm hole in connection with black holes, and also curved space. Can you tell me anything about these?*

Sure, it's simple, just consider the maximally extended Kerr geometry using Kruskal coordinates in....

*Ah, excuse me, again, but I'm having a little trouble following you.*

Oh? Sorry. Well, let's go back to using our imagination. You game?

*You bet.*

Okay, imagine a horizontal drum head, and let's make the drum head very large. Now consider that the drum head is really very thin, or that it has no thickness at all, so that we have a flat, two-dimensional surface.

*By horizontal you mean parallel with the ground?*

Yes, and what happens if you roll a marble on this surface.

*It will roll along in a straight line?*

Absolutely. Now, imagine further that the drum head is elastic, so that you could push it down in the middle, say you stand on it, but it only curves near the middle where you are standing, and becomes flat again far away from you.

*I'm still with you.*

This is just how matter curves space. You can easily imagine the two-dimensional curved space because you have the luxury of being able to embed it in a three-dimensional space. However, matter curves the three-dimensional space, but you cannot visualize that because you are not able to visualize four-dimensional space. That is why I wanted to use the blackboard.

*I follow your analogy. In fact, the heavier I am, the more I would curve the membrane, so the more mass there is, the more the space is curved. Am I right?*

Very well put, and if you were extremely massive, you would make a very deep and narrow tube on that membrane, and you can imagine that if I rolled a marble too close to you, it would fall down into the tube and never be able to get out. That point of no return is...

*The event horizon! Thank you for explaining all this to me, but what about the worm hole?*

I'm afraid that this is really a mathematical discovery, and I gather you would rather not see the math?

*If possible.*

Then let me just say when space is described using coordinates, you are free to use many different coordinate systems. The discovery was, that in a particular coordinate system, all of the space you visualize outside of a black hole is really only half of the entire space. You just don't see the other half.

*I guess I don't.*

Imagine you have a piece of paper and cover the entire piece with coordinates, simply horizontal and vertical lines, like a piece of graph paper. Now suppose you discover it was folded in half, and there were another entire side that you left blank. In other words, your original coordinates were poorly chosen, they only covered half the space. When you visualize the space outside of a black hole, you are only seeing half the allowed space.

Well, now imagine another drum head, just like yours, only very far away. It too has a very massive object in the center, so the center tube is very narrow, and an event horizon is formed there also. Now imagine that these two very thin tubes are connected! That connection is the Einstein-Rosen bridge, or throat, or worm hole. The other black hole is like the other half of the page you missed.

*So every time we see one black hole we should see two?*

No, they may not be near each other. The other black hole can be anywhere, it can be in a different galaxy.

*But you must be connected to another black hole, wherever it is.*

I was, but not any more. This worm hole is a dynamic structure. It stretches out and pinches off — mine pinched off long ago.

*Well, this has been a very illuminating interview. I am beginning to see some of those visitors you mentioned, they go in but they don't come out.*

I'm afraid so.

*And you are growing all the time?*

I'm afraid so.

*Wow, look at the time. I have to be off to my next interview. Thanks very much for the interview.*

## 0.5 Interview with a uranium atom

*Good evening. Could you begin by telling us how you were formed?*
I was born in a supernova explosion.

*A supernova occurs when a star collapses under its own weight, squeezing matter into an unimaginably high density?*
Yes. During that instant matter is squeezed to an extraordinarily high density, huge numbers of protons and neutrons are pressed together and virtually all of the elements are created, including the heavy elements.

*This is essentially the process described in my interview with the black hole?*
Yes, but they are so mathematical I'm surprised you were able to get her to say anything useful, it's Greek to me. You did a good job.

*Thanks, anyway, you were created in a supernova explosion, traveled here, and became part of the Earth?*
Eventually, yes.

*Eventually?*
To be honest, I started out here as part of the leftover rubble in your solar system.

*Rubble?*
I became part of a small asteroid.

*So you were in the asteroid belt, how did you get here?*
I was not actually part of the asteroid belt, which lies between Mars

and Jupiter. I was one of a smaller group, what you now call the Near Earth Asteroids.

*How near?*
Well, most of us were between one and two AUs from the Sun.

*An AU is an astronomical unit?*
Yes, the distance from the Earth to the Sun.

*So what made you decide to come here?*
The invitation was extended by a very rude comet. Bent on self-destruction, and zooming toward the Sun at an ungodly speed, it whooshed so close to me it perturbed my orbit, and I began oscillate, so to speak. At one point I became a little too close to Earth and began my kamikaze journey. The comet got too close to the Sun and — goodbye comet. Served it right.

*When was that?*
I'm not sure, eight, nine hundred million years ago.

*Am I detecting a little jealously between asteroids and comets.*
I wasn't jealous, but those comets sure think they're superior, strutting around like a peacock, spreading their tail a million miles into space.

*Since you were part of an asteroid, and apparently had a close encounter with a comet, perhaps you would be willing to explain the differences between comets and asteroids.*
Well, it's been a while, but let me see. Most asteroids are solid mineral type objects, some are mostly carbon, you call them carbonaceous, some are more iron, but essentially made of solid stuff, not like those flimsy show off comets. Most asteroids stay in nearly circular orbits, although the Near Earth asteroids tend to have slightly more eccentric orbits.

*Eccentric orbits are elliptical in shape?*
Yes, the longer and flatter the ellipse is, the higher the eccentricity. Comet orbits have very high eccentricity. Some can go from ten to thirty AUs and zoom in, just grazing the Sun, some go out much further. As far as I'm concerned they're pests of the solar system.

*Are there other differences between comets and asteroids?*

Well, as I said, asteroids are pretty solid, but comets are just dirty old snowballs.

*Indeed?*

They are mostly ice, water ice, ammonia and carbon dioxide ices, with some minerals mixed in. When they are far out everything remains frozen and they look a lot like an asteroid. But as they get near the Sun the ices melt and sublime...

*Sublime?*

Change from a solid to a gas. As the particles boil off they shoot out into space, but since they are still in orbit too, they form a gigantic tail.

*I see. I am sorry to stray off the subject, but before I ask you more personal questions, could you briefly mention your arrival here?*

I must admit I was taken by surprise. After billions of years of orbiting your star, I found myself accelerating toward your planet and began to wonder what my new home would be like. The first thing we noticed was that we were getting hot on the outside, and for the first time in my life we became a spectacular sight, racing through your atmosphere at over 20,000 miles per hour, we lit up the sky like the last fireworks on the fourth of July. But that was nothing compared to what happened when we hit.

*How big were you?*

About a kilometer across. A mere speck in comparison to the size of your planet, so I was surprised by the ruckus we kicked up. We hit in what you now call Canada, and you can imagine what happens when $10^{21}$ joules of kinetic energy are transformed into heat and shock waves.

*I wish I could, would you elaborate?*

As soon as I hit the ground the enormous kinetic energy went into melting and vaporizing the asteroid along with some of the surrounding Earth. The material, partially molten, propagated outward in a fiery wave. The enormous amount of heat raised the air temperature to thousands of degrees for miles around, setting off a fire storm that burned everything for hundreds of miles. All this happened in the first few seconds, and was really the least of it. A large portion of the remains of the explosion entered the atmosphere as tiny particles, and the smoke from the spreading fires changed the atmosphere from nitrogen and oxygen to nitrogen and carbon dioxide. All, or most, of the life was obliterated by that and what followed.

*What followed?*

The dark and murky atmosphere blocked the sunlight and for about forty years the Earth got really cold. It was not what you call an ice age, but ice was almost everywhere. Finally though, as the particles and soot settled out of the atmosphere, the carbon dioxide captured a large percentage of the reflected infrared radiation.

*You mean there was a greenhouse effect.*

Big time. The entire Earth was scorched after that, where there was once ice you could fry an egg, if you could find one. But, as you know, time heals many wounds. The Earth was never the same, but it came around to a nitrogen-oxygen atmosphere again, and life, as persistent as those comets, began again.

*That's an extraordinary story, do you think it will ever happen again?*

Well, you can't trust a comet. Any one of those buzzards could knock an asteroid off course, and they're ornery enough to crash right in themselves. As you know, that's where you got your oceans. I should add that you are hit all the time, many times a day, but they're mostly small little specks.

*I see. Thank you for your explanations about comets and asteroids, but I would like to get back on target.*

I'm ready.

*I was wondering if you can tell me about something. I have heard of U235 and U238, these are both uranium atoms?*

Yes, I am U238. I have 92 protons and 146 neutrons, and of course 92 electrons, but my nucleus, the neutrons and protons, number 238. If I had three less neutrons, I would be U235. My siblings, or isotopes as you call them, range from U232 to U238.

*So the only difference between U238 and U235 is three neutrons.*

Yes, but that makes all the difference in the world.

*How so?*

Well, you should know that I am unstable.

*Yes, I planned to bring that subject up later.*

Let me address both issues at once, and, for the record, I'll be talking about big atoms. The thing to remember is that protons just cannot stand to be near other protons.

*You are saying that like charges repel?*

Yes. The more protons the nucleus has the more the animosity builds up, and sooner or later one or more will break out.

*That's radioactivity?*

Precisely, sooner or later I'm doomed to decay into smaller atoms. But I'm lucky, I have three extra neutrons. The neutrons essentially mollify the protons, keeping them apart, giving them a little more elbow room. Since I have three more neutrons than U238, it is a little less volatile in my nucleus, and I have a much longer life expectancy.

*How long will you live?*

My half-life is...

*Would refresh my memory about half-life?*

Sure. Suppose you had a thousand objects in a room, and one week later there were only 500 left. After one more week there were only 250, and after another week there were 125. Also, as far as you can tell they are all identical, and you have no way of knowing which will disappear.

*Then these objects would have a half-life of one week?*

Precisely. The part that gnaws away at your stomach, if you happen to be a uranium atom, is that you can't tell when your time is up. I could go anytime, I can only tell you about how things look on average when you have a lot of us.

*I see, so what is your half-life?*

My half-life is nearly five billion years, but U235's half-life is only a little more than one tenth of that, which is part of the reason it is so rare.

*This brings me to another area I wanted to venture into.*
Be my guest.

*As you must know, uranium is used to make atomic bombs. I see you are uncomfortable with this subject, but can you talk about it a little?*
Well, let me give you a little history. I was dug up during the Manhattan project, during a war that A, you seemed intent on killing yourselves in, and B, I wanted no part of. Anyway, for every 997 of me, there are about three U235 atoms, which is what you use in your bombs. So you go to great lengths to refine the ore, first to get uranium, then to get U235. Once they obtained pure, or nearly pure U235, it goes into the bomb.

*Can you tell us anything about that?*
I can give you a firsthand account.

*You mean, you were part of a bomb? I thought only U235 was used.*
I was an impurity. We all heard about Hiroshima, and then Nagaski, and were bewildered. None of us ever witnessed such large scale annihilation of so many of us. We were shocked and horrified, and for the first time I wished I could be back on an asteroid. I understand the physics of atomic bombs well enough, but I'll never understand how you can destroy yourselves with such passion. Sometimes it appears that your prime objective is to rid the Universe of life. It was very depressing to be forced into such a terrible scheme, and to lose so many.

*Yes, we are doing everything we can to avert another such a destruction.*
You think so?

*Yes, I do.*
I don't know who is worse off, you or me.

*What do you mean?*
Well, I know that I could buy the farm at any time, but with us, unwilling as we are, at your disposal, your entire race can go any moment.

*But...*
I was there, I know.

*Know what?*

You sure you want to know?

*Absolutely.*

By dumb luck I avoided being used during that war. But all left over radioactive material is kept and refined, and I became part of a bomb made years later. I was kept in storage for a few years, and then loaded in a bomber. At first I thought it was some sort of test run, but when I was activated, the only thing between me and Armageddon was about 5,000 feet.

*You're scaring me, what do you mean?*

I was totally armed, as you call it, and all set to be released from the bomb bay. At that point, the altimeter takes over, and detonation occurs at maximum destruction altitude. We were at 5000 feet over the MDA. There was a stand down only minutes before we were to be released.

*This is incredible, when was this?*

Are you sure you want to know?

*Maybe not. I really didn't mean to get into political issues.*

Uranium and politics go together like sodium and chlorine.

*I see. Nevertheless, I was wondering if you would tell us a little about how an atomic bomb works.*

Well, you know that at any moment I can split apart. When I do I give off energy, that is the energy you want to harness, or use to do your dirty work. However, my demise can also be helped along. One way is to send a neutron into me. If it has the right speed, that will certainly induce me to split apart. The key to the whole process is that when I do split apart I give off a couple of extra neutrons. They, in turn, can cause other nuclei to undergo induced fission, and they trigger more fission, and so on. Within a split second, virtually all of the atoms have undergone fission.

*We call that a chain reaction. But how do you start it?*

We don't, you do.

*Sorry, how does it begin?*

Critical mass. Once you put together about ten kilograms of highly refined U235, there will be enough neutrons to begin the runaway process of induced decay. In a bomb, you keep two pieces, each below critical mass, separated. To detonate, you slam them together, usually using a chemical explosive, and reach critical mass. Then, whamo.

*That's all?*
That's all.

*How did you ever escape from the atomic bomb?*
To be honest, I was not part of an atomic bomb, I was part of something much worse, a hydrogen bomb.

*You'll have to excuse me, I can't get over this. Would you explain the difference between an atomic bomb and a hydrogen bomb?*
Keeping it simple, think of a hydrogen bomb as an atomic bomb immersed in hydrogen. When the atomic bomb is detonated, the the hydrogen is heated up to millions of degrees, which means the protons have enough energy to slam into each other. Of course neutrons are important in this process also, but the overall idea is that the hydrogen undergoes fusion into helium, just like in the center of a star, and that fusion gives off energy.

*Hold on a second.*
I'd be glad to.

*You said that when uranium undergoes fission, splitting apart, it gives off energy, and when hydrogen undergoes fusion, coming together, it gives off energy. This seems to good too be true.*
Or too bad to be true.

*Yes, now that you mention it. Can you really have it both ways?*
Yes, remember, when I was talking about fission I was talking about heavy atoms. Fusion occurs with light elements, hydrogen to helium, helium to carbon, and so on, right up to iron. Anything heavier than iron likes to undergo fission, and anything lighter likes fusion.

*Why is that?*
Well, think of a nucleus as a civil war battleground. The protons are pushing away from each other due to the electric repulsion, while

the nuclear force holds everything together. The electric field remains strong over larger distances, but the nuclear force, while stronger at close range, becomes weaker at large distances. For example, in me, the proton on one side can barely feel the nuclear attraction due to a particle on the other side, but it feels the sting of electric repulsion very well. Thus, if you make the atom big enough, the electric repulsion over powers the nuclear attraction. Iron is the middle point, the most stable atom in the Universe.

*I see, but how did you ever escape from the hydrogen bomb?*

Routine inspections revealed fissures in the outer casing. The device was disassembled and, at the same time, improvements, as you would call them, were made to the atomic bomb. I escaped during that process, and although there was some attempt made to keep me entombed in that facility, I escaped into your atmosphere, and eventually agreed to your interview.

*Fascinating. What do you propose to do in the future?*

Of course, I'm always hoping there is a future, the trouble is, I don't know if my demise will come about from natural or unnatural causes.

*I can assure you that many people are working hard so that you will not suffer an unnatural demise.*

But many other people are working in just the opposite direction. That's all right, I knew what I was in for a long time ago. To be honest, I feel lucky to have been born in the first place, and even luckier for those three extra...

*Oh dear, uranium, where did you go?*

## 0.6 Interview with a fermion and a boson

*I appreciate that you both agreed to a joint interview...*

The boson interjected,

"I feel compelled to state that, no, I must say I even object to the fact that the fermion's name appears first, that is, before mine, in your title."

*I'm sorry, I chose the order randomly, I would be happy to commute the order. Perhaps, though, we should begin. Would you tell us the difference between a boson and a fermion.*

Before the fermion could get a word in the boson answered,

"I really don't know why she's here, bosons are where the action is. We are dynamic, we produce interactions, we make things happen. These dull fermions would mope around in a pretty dull world without us."

The fermion waited patiently but finally said,

"I will explain the differences between us. First, all particles ever observed are either fermions or bosons."

The boson interjected,

"See? She put the fermion first. All right, okay, go ahead."

The fermion continued,

"Particles have spin, like the Earth spinning on its axis..."

The boson interupted,

"No, it is not like the Earth spinning on its axis, particle spin does not arise from motion, we are endowed with spin like we are endowed with mass, or charge."

The fermion replied,

"He's right, I was making an analogy. Anyway, spin is measured in units of $\hbar$, which is Planck's constant $h$ divided by $2\pi$. Any particle that has one half, or three halves, etc., $\hbar$ is called a fermion. Any particle that has an integer number of, including zero, units of $\hbar$ is called a boson. Sometimes we simply say fermions are spin one half particles while bosons are integer spin particles."

The boson added,

"Wrong again. These sweeping generalizations are typical for a fermion. The spin she was talking about is the component of spin along some axis, we usually call the $z$ component.

*With this proviso the fermion is correct?*

"Yes," the boson admitted.

*Could you give me some examples of fer... I mean bosons and fermions?"*

The boson answered,

"Sure, a photon, a particle of light, has spin one, and is a boson. The $W$ and $Z$ particles which account for the weak nuclear interactions are bosons, the gluons which account for the force between the quarks have spin one, and of course gravitons of the gravitational field are spin two. All bosons.

The fermion added, with a nonzero trace of anger,

"All the stable, well-known particles are fermions. For example, electrons, protons, neutrons all have spin one half and are therefore fermions. The quarks, which make up the protons and neutrons, have spin one half and are therefore fermions."

The boson interrupted,

"Big deal. Without bosons they would not interact. There would be no force to hold the nucleus together, no electric force to make atoms form, no magnetic force. Nothing. The Universe would be nothing but a bunch of single particles moving in straight lines. There would be no stars, no galaxies, no planets, nothing.

*That would certainly be a dull Universe. What exactly is the role of the bosons, then?*

The boson continued,

"Suppose you have a force between two particles, say two electrons repelling each other. How does this force come about?

*Well, we know like charges repel. As I understand it, the electron creates an electric field, and that electric field exerts the force on the other electron.*

The boson, calming down a little, said,

"Where have you been? That view is older than your tie. What actually happens is this; one electron creates exchange particles, photons, and these are absorbed by the other electron. The exchange of the photons is the fundamental origin of the force between them."

*Do all particles exchange photons?*

The fermion answered,

"No, only particles with electric charge do. However, the quarks that make up the protons and neutrons exchange gluons..."

"Which are bosons," interjected the boson.

The fermion continued,

"Which are bosons. They account for the strong nuclear force."

*Hold on. You are saying that neutrons and protons are held together because the quarks, which make them up, exchange particles called gluons?"*

The fermion edged out the boson saying,

"There is a little more to it, but yes."

*Okay, so the fermions, the electrons, quarks, protons, etc., interact with each other by exchanging bosons. So it seems to me the difference between bosons and fermions is not simply your spin, but you play fundamentally different roles.*

The boson blurted, "And without us you would have a pretty poor world."

The fermion took the higher ground, and ignoring the boson said,

"You are right. It takes both kinds of particles to create the Universe in which we live. Both are required, and both may lead rich lives. I would not like to live in worlds without bosons, nor could I imagine one without fermions.

The boson seemed upset and began a new line of argument,

"Ms. high-and-mighty sounds very egalitarian, but don't believe it! She's a snob, all fermions are snobs."

*A snob?*

The boson was excited again, and continued,

"Ask her to deny it. She won't, she can't. Particles, whether bosons or fermions, are described by being in a particular quantum mechanical state."

*Quantum mechanical state?*

The boson continued,

"All that means is this; by natural law, we are allowed to have only certain allowed energies, certain allowed momenta, and so on. Once you specify what these things are, that is called our quantum mechanical state, or for short, state."

*I see.*

The boson went on,

"Now I ask you, do you know how many particles are allowed in a given state?"

*I'm afraid I don't.*

"Well I'll tell you. If your talking about bosons, there is no limit. As many particles that want to can share the same state. We exclude none. But if you are talking about those snobby fermions, the answer is one and only one. Once a fermion occupies a state, none other is allowed into it. Go ahead, ask her to deny it."

"Of course I don't deny it, it is called the Pauli exclusion principle," countered the fermion, and continued, still ignoring the boson, "that property is precisely what gives the Universe its rich structure. If all electrons, which are fermions, were to occupy the same state, than atoms would be hardly distinguishable and you would not have the wonderful complex structure you see. In fact, bear in mind that since you are very nearly empty space, without this property, that chair could not hold you up. For that matter, your planet could not even exist, at least not in anything like its present form."

"Hmmph," from the boson.

*If I may summarize, might I say that fermions are the brick and bosons the mortar of matter?*

"Well put," echoed both particles, with the final comment coming from the boson,

"And without the mortar you can't build anything, the bricks would lie there in a useless heap."

*Well, this has been a fascinating interview. I would like to thank you both for coming by, and helping me understand your differences.*

## 0.7 Interview with a star

*I should mention to our readers that you are not any star, you are the one we call our Sun. Thank you for agreeing to this interview.*

I am glad to do it. May I suggest you use darker sunglasses?

*Yes, that's better, thanks. I understand you were formed from a great cloud of hydrogen gas about ten billion years ago. Is this correct?*

Yes, thinking back to my dim beginnings I see mostly black, empty space. Hydrogen, a little helium and a sliver of heavier elements were scattered across thousands of light years of space, about as crowded as lakes on a desert.

*Is it true that the gravity created by this matter began to pull you together?*

Yes, at first my atoms felt very gentle tugs, like skiers starting from from the top of a rounded snowcap. They serenely glided toward an unidentified center, happy having acquired a direction in life, but quite ignorant of what was in store.

*Then what?*

This continued for millions of years, but eventually the serenity was lost. Collisions between atoms, which earlier had been a rare occurrence, became the norm. It seemed that the atoms were fighting to find the center, and the once vast cloud was shrinking, by my old standards, to an incredibly small size. Near the center, things were heating up and collisions became so frequent something wonderful happened.

*What was that?*

It began to glow.

*It glowed because it was hot?*
Like a poker in a fire.

*It was hot because of the collisions between the atoms?*
It was hot due to the speed of the atoms. The hotter something is, the faster the atoms, or molecules are moving. Think of heat as an average measure of speed. By the time the atoms and molecules neared the center, not only had they been accelerating for a long time, as the matter bunched together, the gravitational field got stronger, so the particles were really moving along.

*So, like any hot object, it began to glow.*
It was wonderful. Light began to fill the blackness, and all of the nearby atoms and molecules got excited. We saw that the vast cloud had collapsed to a swirling disk, and as the disk continued to shrink, its rotation rate was forced to speed up.

*Is this conservation of angular momentum?*
Yes, like an ice skater pulling in her arms. As she pulls together, she must spin faster. But with me, I could not quite hold myself together, so pieces pulled off and went into orbit around me.

*The planets?*
Yes, and Jupiter took up most of the angular momentum. In fact, this is very common, and at least half of the stars in the galaxy are binary stars for just this reason, something Jupiter seems to know all too well.

*It would seem that you are saying, if there is a star, there is either another star or planets around it. That would make a lot of planets.*
It sure does.

*What happened next?*
The warming glow became increasingly more intense, but gravity gave no quarter. We continued to collapse to a denser and denser ball. Then it began to happen, a flicker here and a flicker there, and we stood in awe as we witnessed a miraculous process begin, a process that caught us totally unawares.

*What was happening?*

We were so hot in the center that the electrons were ripped away from the hydrogen atoms, leaving bare protons. The electrons were in shock, blinded by the light, hovering around trying to grab onto protons, but were continually knocked free. To be honest, we were in a momentary daze as we witnessed all kinds of miracles. Many particles were being created and destroyed from the immense energy associated with the intensely hot plasma, and that is when it began to happen, as I said, first here or there, but then everywhere within the core of the hot ball.

*What?*

Fusion. Hydrogen was being transformed into helium, and with the formation of each helium atom, energy was being released into space. In fact, so much energy was being released and sent outward, away from the center, it created an extremely large pressure, what you call radiation pressure. In a star there is a continual war raging, the inward pull of gravity which wants to see total collapse, against the outward push of radiation pressure, trying to get free. These forces can be very large.

*The force of freedom can be devastatingly large. In your case, how large?*

Large enough to stand up to gravity. This outward radiation pressure ultimately balanced the inward pull of gravity and the peace treaty was signed. We have been in peaceful equilibrium for nearly ten billion years.

*You have had a spectacular birth. It is as though you underwent a metamorphosis, from a black cloud to a shining star.*

Metamorphosis is to nature like sand is to your desert.

*After all that, do you find things getting dull now?*

I have my moods.

*Perhaps I can review your vital statistics?*

By all means.

*You have a mass of* $2 \times 10^{30}$ *kilograms. Can you put that into perspective?*

It's about a 300,000 times the mass of the Earth, or about a thousand times the mass of Jupiter. I am also about one million miles in diameter.

*I've jotted down in my notes that your luminosity is $4 \times 10^{26}$ watts. Can you explain this to me?*

Luminosity is just another word for power. Your 100-watt light bulb, I see you turned it off, has a luminosity of, simply, 100 watts. So I am somewhat more than a trillion times a trillion brighter than that. Your neutrino explained it correctly.

*I see, and you spin on your axis?*

Yes, it takes about a month, a little less along my equator, to make one revolution.

*You said you were very hot, how hot are you?*

In the center I am about 15 million degrees, but on my surface, the part you see, I am about 6,000 degrees.

*All of the energy you emit arises from the conversion of matter into energy, correct?*

Correct.

*So am I right in assuming that you are continually losing mass, getting smaller every day?*

Yes, but do not worry about it. I am losing about five million tons each second. That's only about 150 trillion tons per year.

*Only!*

Spanning ten millennia, that amounts to less than one billionth of my mass — believe me, I can spare it. In fact, I get a great return on my expenditure. I get to observe all my planets, religiously orbiting me, my wonderful comets and asteroids, and of course, the curious antics on earth — not to mention this interview.

*Your energy is certainly supportive. I understand you also have a magnetic personality.*

Yes, like your planets, most things that spin have a magnetic field.

*Why is that?*

It is not really understood how the magnetic field arises from all of these objects, but for me, like your Earth, the origin of the magnetic field is a kind of dynamo effect. The rotation essentially causes tremendous electrical currents to flow in great loops, and since an electric current makes a magnetic field, I have a magnetic field.

*That's all there is to it?*

Well, there are some twists to the story. As I alluded to earlier, I undergo differential rotation, which simply means that the material on my equator goes a little faster than that on my poles.

*Yes, I remember.*

This uneven flow twists up the magnetic field lines, and sometimes they get into a terrible tangle.

*Sounds like a bad dance.*

In a way it is. The magnetic field lines sometimes get trapped in certain spots which sashay across my surface. Imagine great handles, like on your coffee mug, attached to my surface.

*Like handles on a suitcase?*

Yes, but the handles are bigger than your planet. These handles are intense magnetic field lines, coming out at one spot, and delving back in at another spot. These strong magnetic fields deflect some of the energy that wells up from my interior, making these so called sunspots about one thousand degrees cooler than the surrounding material.

*They appear darker because they are cooler?*

Yes, the hotter something is the more energy it gives off.

*So the sunspots are not really black, as they appear?*

No, if you could block out the rest of me, they would seem very bright.

*I was wondering about something you said earlier.*

Yes.

*You are changing hydrogen to helium, which is the source of your energy, at an incredible rate.*

It's credible.

*Well, I meant, to us, a very fast rate.*

Okay.

*So, what happens when you are totally helium?*

Do you remember that I said there is a continual war raging within me, gravity pulling in against radiation pressure pushing out?

*Yes.*

When hydrogen burning, as you like to call it, stops, gravity smells victory and closes in for the kill — I start to collapse.

*How unfortunate for you.*

No, it's great for me. Greedy gravity gets foiled by its own doing.

*What do you mean?*

Press your hands together, hard, and rub them back and forth.

*What?*

Go ahead, four or five times is plenty, that's it. What do you notice.

*The're warm, no, the're hot.*

Friction heats things up. Same idea with me, as greedy gravity starts to push all the helium atoms together, they get hot.

*I don't blame them.*

Now, let me be a little more detailed. As time goes on, my center becomes solid helium, surrounded by hydrogen, still undergoing fusion. The helium begins to collapse because, as I explained, gravity is there waiting. However, that heats things up, and the hydrogen on the surface starts burning faster than ever.

*By burning, you mean undergoing fusion?*

Yes, it gets so hot the radiation pressure overpowers gravity.

*What happens next?*

The hydrogen expands way beyond my normal surface. As it expands it cools.

*Is this the same principle as in our air conditioners, expanding gases cool down?*

Precisely. It goes from being white hot to red hot, down to about 3,000 degrees. At that point I am very big and rather cool — on the outside. You have named those stars red giants.

*It would be interesting to see a red giant.*

Take a gander at my old friend Betelgeuse, in Orion, he's a red giant. You should be able to see that it is really red.

*I'll check it out. Just how big will you get?*

Big, but I'll get bigger.

*What do you mean?*

Remember all that heat from rubbing your hands together? That continues in my core until the helium gets so hot it undergoes fusion to form carbon. That sends another heat wave through the outer hydrogen, making a red supergiant. At that stage, even your planet, I am very sorry to announce, will be within my surface.

*That won't be too soon, will it?*

About five billion years from now, give or take.

*That's comforting. So does this process keep continuing, helium to carbon, carbon to...*

No.

*No?*

Not for me. I end up as a solid carbon sphere, very hot at first, but as I continue to emit radiation I cool down. The outer layers of hydrogen and helium keep getting further away, and actually break free of gravity. For a time, you can see a star, that's the carbon remnant, surrounded by beautiful clouds. You call these planetary nebulae, although that name makes no sense to me. Eventually this material goes into interstellar space, and some day may help form a new star, all over again.

*So, in the end, you do lose quite a bit of mass.*

Yes, at the end, but it's like sending your children out into the world. You just hope they get their turn to shine.

*And you, as a white dwarf?*

At first I shine pretty brightly, even though I am very small, which is why you call me a white dwarf. Since this energy is only stored as heat, and I cannot generate any more, I cool down fast. In a few million years I barely shine at all.

*Will you still be a white dwarf?*

I will stay the same size, but get cooler and cooler, and therefore dimmer and dimmer. Eventually, I will not emit enough light to be seen, and you will call be a black dwarf.

*Sounds like a dismal end to such an exciting life. You just fade away?*

Like an old soldier.

*Now I understand something the carbon atom said.*

What is that?

*She said, roughly, "as my star cooled, I came to the realization that I would be forever trapped inside a giant, inert star made of nothing but carbon copies of me."*

Yes, she was in the white dwarf stage, facing the black dwarf fate, but was saved by her companion star, and underwent a supernova.

*Can you explain that?*

I see you have an interview scheduled with a neutron star. May I suggest you ask her?

*Yes, I will. May I ask you one final question?*

Yes, go ahead.

*Can you tell me which is your favorite planet?*

Well, Mercury is of course the closest to me, but I suppose that is more of a matter of location than anything else. It is too small to have an atmosphere so it feels my blistering rays directly, it really feels the heat. I like Venus all right, which is about the size of Earth but very hot by your standards, keeping herself wrapped in a thick insulation, an atmosphere of carbon dioxide. Let me see, what comes next...

*Earth.*

Oh yes, Earth is very special, but can be disappointing.

*How so?*

Well, you seem to spoil a lot of things that took, not only me, but almost the entire universe to make.

*It seems we would be hardly capable of such things.*

You do it with a vengeance.

*Can you be more specific?*

Well, it took thousands of millions of years, and countless comets, one of my most beautiful sights, to sacrifice themselves by plunging into your planet to make your rivers and oceans. They gave you life and beauty. Your mission seems to be to change their water into some indefinable sludge containing more contaminants than stars in the sky.

*We're working on cleaning things up.*

It took me millions of years to build up an oil supply, too many trees to count grew tall and spent eons underground, decomposing, to give you oil. Now you squander it, as though it is your inalienable right to burn fuel.

*We're working on that too, but these are things that came from our solar system. You mentioned the entire universe?*

Your entire earth, and the comets too, came from stars that exploded in the remote past. I was thinking of your uranium atom, I never really thought about it from that perspective. It takes in incredible amount of matter to make a star, and almost an eternity for it to go through its life cycle. At the end, when it is in its death throws, it creates a tiny amount of uranium and plutonium, and, with its dying gasp, ejects it into space. Free of gravity, I might add.

*Yes, we are able to use these materials.*

All too quickly. You gather them up, obtain critical mass, and in one thunderous millisecond they are gone, destroying your humanity while leaving behind further contaminants to your planet. In virtually no time you destroy the entire fortune of billions of years, taking as much life with it as you can. It is a sad thing to see.

*So Earth is not high on your list?*

I didn't say that. You have had wonderful success in understanding me and my constituents, you have had great thinkers and philosophers, your art and music is unexcelled. With all the truly magnificent accomplishments you have managed, and in a short time, it is disappointing to see that you have such a dark side.

*I see, and the other planets?*

I like Mars, a real trooper, hanging in there with its two tiny moons and an atmosphere as rare as gold. Jupiter is special, with all those moons, and Saturn, with the rings, is a real delight. Uranus and Neptune keep to themselves, but you have to admire Pluto, way out there, plodding along, colder than ice.

*But no favorite?*
Yes, I do.

*Which one?*
My favorite planet is... Yikes, look at the time, it is time for me to set. Sorry, gotta go.

*Thank you for the...*

### 0.8 Interview with a Wimp

*Thank you for granting me this interview, I realize that many people do not even believe that you exist.*

One of the reasons I agreed to this interview was to clear up that and this business about my name. I'd rather you not call me a Wimp. I happen to be a neutralino.

*I meant no disrespect, I assure you. Maybe it'd be best if you begin by explaining what a Wimp is, and what what a neutralino is.*

A Wimp is a weakly interacting massive particle.

*Ahh-huh... could you explain a little further?*

Okay. As you know, there are four fundamental forces in nature, gravity and electromagnetism are the most evident to you, but there are two nuclear forces as well, the strong nuclear force and the weak nuclear force.

*Yes, I just had an interesting exchange with a fermion and a boson about this.*

I should point out that there are unified theories in which all of these, except gravity, are really different manifestations of the same thing. For example, the weak nuclear force and the electromagnetic force can be viewed as different aspects of the same fundamental force.

*Is that the electro-weak theory?*

Yes, and it predicts the existence of the W and Z particles, which your electron friend mentioned.

*I have a pretty good idea of what gravity is, and am somewhat familiar with electricity and magnetism, but these other two forces, the weak and strong nuclear force, can you elaborate here?*

Yes. Your carbon atom mentioned fusion, and the uranium atom talked a little more about the force that holds the nucleus together. This is the strong nuclear force. It is stronger than the electric repulsion of the proton — it is the strongest force in nature.

*Wait a minute. In my interview with a black hole, she said gravity was the strongest force.*

They always say that. Technically speaking, yes, she's right. That is because matter can muster itself together in large quantities, and the gravitational force is a long range force. The nuclear force dies out very quickly, as the uranium atom explained, and cannot compete with gravity on the large scale. On a "per particle" basis, gravity is so weak we ignore it!

*Okay, thanks for clearing that up. So, the strong nuclear force holds protons and neutrons together in nuclei. What about the weak nuclear force?*

The weak nuclear force, or weak force for short, is much weaker than all forces except gravity. Nevertheless it is important. While the strong force acts between nucleons...

*Nucleons?*

A nucleon is a term used to refer to either a proton or a neutron. As I was saying, while the strong force acts between nucleons, the weak force acts between an electron and a nucleon.

*So the electron is immune to the strong nuclear force, but feels the weak nuclear force?*

Precisely. Any particle that happens to feel the weak force, and not the strong force, is said to be weakly interacting. It is also called a lepton.

*So you feel the weak force and therefore are a weakly interacting massive particle. But what about the word massive in there?*

I'll remind you that, this interview aside, I have never been detected. I am a purely theoretical prediction, and you have not had the opportunity to measure my mass. You expect me to be between ten and one thousand times as massive as a proton, perhaps more.

*Wow, that's heavy. I understand what a Wi... I, mean, a weakly interacting massive particle is, but, I am confused about what a neutralino is. Could you explain this a little?*

Sure, its a super story, but we must take a detour first. You sure you want to hear it?

*Absolutely.*

Okay, you must remember that argument between your boson and fermion. I don't mean to laugh, but they really went after each other.

*You should see my original notes.*

They gave you the standard view; bosons are bosons and fermions are fermions, period. However, there is another way to view nature, which is still in its theoretical stages. Your two friends would die if they were here now. In any case, they each shared the underlying notion that a boson cannot turn into a fermion and a fermion cannot turn into a boson.

*Yes, I gathered that.*

This notion seems valid on experimental grounds, but has been called into question in the theoretical arena. Certain nasty mathematical problems can be avoided if one allows for the possibility that these particles can transform into one another. Many people think it is also aesthetically pleasing.

*So an electron could change into a photon?*

No, not quite, you would lose the negative charge of the electron, but you are on the right track. Let me put it this way. We view fundamental particles and their interactions in terms of symmetry.

*Like if I write something in wet ink on a piece of paper and then fold it?*

No, not at all like that. The idea really started with the neutron and proton. As far as the strong force goes, these particles are the same, and therefore we began to consider them two different states of the same particle, like Dr. Jekyll and Mr. Hyde, same person, but he can adopt either identity. This is a symmetry, sometimes called a particle symmetry. Now that I think about it, it is a little like your folded paper analogy.

*Oh? You just convinced me it was quite different.*

Well, take your paper and rotate it 180 degrees. It appears to be the same, right? In fact, if you fell asleep for a second you could not tell if someone sneaked into the room and rotated it, or if it was left alone, right?

*Yes, this is true.*

Same idea in particle physics, if you interchange a neutron and and proton in a nucleus, for example, except for the change in the charge, you would have the same thing. This is the symmetry we are talking about.

*I see, are there other symmetries?*

You have asked the million dollar question. The answer is yes, but you are struggling to find out precisely what they are. For example, we mentioned the electro-weak theory. In that theory, the electron and the neutrino, which appear to be quite different, are considered to be different states of the same particle.

*I had an interview scheduled with a neutrino, but she was hard to stop. I think I have another scheduled for next week.*

Good, she can give you more details about their differences, or similarities. In your very successful standard model, you view the quarks, the electrons, and the neutrino in this fashion.

*So, all of the fermions are considered different states of the same particle?*

In a way, yes, but do not think of a particle as having a strict identity.

*This is a little confusing.*

Okay, pretend you are dreaming about a piece of fruit. At one point this piece of fruit is an apple, but a second later it is an orange, and then it is a banana. As a matter of fact, each time you look at it it is one of these fruits, but you are never sure which it is until you look at it.

*This I understand.*

If you wanted to describe the fruit, you might say it is a piece of {apple — banana — orange}, or apbanor for short.

*Okay.*

Then you would make a rule, when ever I look at the apbanor I see an apple, or a banana, or an orange. That would be a mathematical way to describe how things work in your dream. You might even assign a probability for finding each particular fruit.

*Yes, but that is just a dream.*

The crazy thing is, that is not just a dream. That is way things work on the subatomic scale. The symmetry we are discussing is the fact the nature mixes up the apple, banana, and orange, or in reality, the quark, the electron, the neutrino, etc, in just this fashion.

*Fascinating.*

Well, that is the end of the detour. Now I can begin to explain what a neutralino is.

*Please do.*

As I hope you appreciate, our view of the world, or the fundamental particles that make up the world, is based on symmetry principles that, in essence, mix up different particles.

*Yes, I see. However, it occurs to me that the symmetry you described mixes together only fermions. Electrons, quarks, and neutrinos are all fermions.*

Precisely. Then the idea came along, why be so restrictive? Let us consider a more general symmetry that can change fermions into bosons?

*Why not?*

Well, the main reason is that such a thing has not been observed. In all your experiments you have never seen it happen.

*Oh, then perhaps I should have asked, why consider such a thing?*

As I said before, certain nasty mathematical and unphysical problems disappear. This symmetry is called supersymmetry, and it predicts, for every particle we know of, a super partner. You can think of the supersymmetry as an operation that can transform a fermion into a boson and a boson into a fermion.

*So, an electron, which is a fermion, has a superpartner that is a boson?*

Yes, it is the selectron. We get the name but putting an s in front of the fermionic names. So electrons, neutrinos, and quarks have superpartners, which are bosons, called selectrons, sneutrinos, and squarks.

*And do bosons have superpartners?*

Yes, you get the name by adding "ino" to the bosonic name. The superpartner of the photon is the photino, the superpartner of the gluon is the gluino, and so on.

*And you?*

Finally. There are several superpartners of bosons that can be mixed together. I mentioned the photino. There is also the Z boson, remember? The superpartner is the Z-ino. There is also a Higgsino. I represent any combination of these.

*So you are a fermion?*

I am.

*And none of the ino particles, or selectrons, or any of those has been observed?*

No, none.

*Then, I must ask, is supersymmetry real, or just the dream of theoretical physics?*

As you know, although many of us have agreed to your interviews, there are some questions we felt were of too personal a nature, and we decided we should not answer. In addition, we believe that we carry certain confidences that we, as a body, are not at liberty to divulge. In other words, you are allowed to probe only so deeply, but beyond that, well, we have our rules too.

*Okay, but let me try this. Of all the putative superpartners you mentioned, you are most sought after. Can you explain this?*

Yes, it's because I am the lightest, least massive. You see, any superparticle that is more massive than I am, can decay into lighter particles. The buck stops with me. I cannot decay into anything because there is no superparticle less massive than I am.

*I guess you are lucky.*

Very.

*There is one more thing, if you don't mind.*

Not in the least.

*You mentioned the standard model, could you amplify on that a bit?*

It is the most fundamental, correct theory you have. It describes how all of the fundamental particles interact, and even predicted the existence of particles that were subsequently observed. Until recently, every experiment ever performed added nothing but brick and mortar to this theory.

*What do you mean?*

Virtually every experiment you performed gave data that was perfectly explained by the theory; any test you could dream up was passed with flying colors.

*Then the theory is correct?*

It seemed so, until recently.

*What happened?*

I'm afraid I haven't heard the details, but it has something to do with the anomalous magnetic dipole moment of the muon?

*Could you explain a little further?*

I think you should ask the muon directly, as I said, the details are hazy to me.

*Okay, I'll do that, I have an interview scheduled with a muon, and let me thank you for such a delightful interview, good luck.*

Thank you, and good luck to you.

## 0.9 Interview with a comet

*Thank you for stopping by, could you tell us something about yourself?*

Measuring 20 miles across, I am a relatively large comet, made of frozen water and carbon dioxide, simple carbon compounds, and a few other earthly ingredients.

*Have you been here before?*

Yes, during my last visit, I noticed pyramids being built and rudimentary metallurgical processes get underway. As I began heading away from the Earth Pythagoras discovered that the frequency of a musical tone is inversely proportional to the wavelength. Inspired by his results and mathematical simplicity, he conjectured that the planets should be equally harmonious, and therefore fall into integer distances from the Sun. Although I knew he was wrong, I was pleased to see incipient theoretical attempts to understand the heavens.

*I see that you have an interest in our science. Is there anything else that struck you during your last visit?*

I still was traveling away when Democritus postulated the existence of atoms, and during the next two centuries when Archimedes presented remarkable results on mechanics and buoyant forces. About a century later, the first century BC, Ptolemy performed a series of studies in reflection and refraction, and produced numerical tables, proving, among other things, that the index of refraction of glass is greater than that of water.

*Index of refraction is the amount a substance can bend light?*

Yes, that is one way of looking at it. Ptolemy also assumed all things orbit the Earth, but even though he was wrong on that, his cosmological work was useful for over one thousand years. Had I free will, these events alone would have caused me to return, but the incorporeal fingers of gravity still had me in their grip, and back to Earth I was bound, nevertheless.

*Where were you at this point?*

At this stage I was a little over 500 AU from Earth. This part of my orbit, when I am farthest from the Sun is called aphelion. The Sun appeared about 4 millionths as bright as it does on Earth, appearing like your 100-watt bulb 100 feet away, and I was crawling along at about 200 miles per hour. At that rate it would have taken me 27,000 years to get back to Earth, but even though my acceleration was only 2 billionths of that of the apple that fell on Newton, it would increase — and so would my speed. My only source of heat is the Sun, and at aphelion I am only 15 degrees on the absolute scale, which, on the Fahrenheit scale, is 433 degrees below zero!

*Sounds like you really cooled your heels.*

They were cooled even further by the surprising decline in science. It seemed to me that the great achievements in science in the preceding centuries were being lost — and I was doomed to remain incomprehensible for ever. It took me 13 centuries to edge in from aphelion to 400 AU. Still being about ten times further from the Sun than Pluto, I was heartened to see the School of Science open in Baghdad, and to see that it was able to save and translate some of the earlier scientific work of the Greeks. With the development of algebra there, although the Sun was still 6 million times dimmer than it appears on Earth, things were getting brighter for me. My temperature only went up a few degrees, but I was now traveling at 2,300 miles per hour.

*I did not realize how long you spend in deep space. I would imagine you enjoy the trip in.*

Yes, from 1600 to 1700 I traveled 3 billion miles, and these were 3 of the happiest billion miles I ever spent. Galileo used the recently invented telescope to discover four moons of Jupiter, and more importantly, that Venus exhibited phases like the Moon. From this he rightly concluded that it revolved around the Sun. Kepler, using the many precise measurements of the positions of the planets made by Tycho Brahe in his elaborate observatory, deduced that the orbits were ellipses, and pronounced his famous law that the period squared is proportional to the semi-major axis cubed. I was amused at Kepler's earlier result that explained why there were precisely 6 planets due to the fact that there are 5 regular solids, and from this he worked out 6 allowed orbits. Unlike my last Earth flyby, this time I felt like progress in physics might keep pace with my continued acceleration.

*What kinds of physics did you see develop?*

The 18th century was kicked off with Newton's (finally) publishing *Opticks.* Newton believed that light consisted of particles, although he further believed that these "corpuscles" would vibrate, so it took a while before the work of Young, who postulated that light consists of waves, was accepted. Fahrenheit developed the scale that is still used in America, and experiments on static electricity were being performed. Halley used Newton's theory of gravity to correctly predict that one of my sisters would return in 1758, but it was a few days late due the effects of Jupiter, which also had an important effect on me. Herschel discovered Uranus, which was earthshaking because all of the other planets were known before recorded history, and opened the door for the question "Are there other planets?" By the end of this century Volta put together a series of alternating copper and zinc discs, separated by moist pasteboard — the first battery. By then I was ten times further from the Sun than Uranus, but I was traveling at nearly 5,000 mph.

*This brought you to the 19th century?*

Yes. During the 19th century I traveled nearly five billion miles. I saw John Adams and Urbain LeVerrier struggle over the orbit of Uranus.

*What was their struggle about?*

Although Kepler observed that orbits of the other planets were ellipses, and although Newton's theory predicted elliptical orbits, the Uranus orbit was not quite an ellipse. This problem really stirred things up.

*What do you mean?*

You were stumped, nobody could understand why the orbit was wrong.

*What was the solution?*

Well, some people pointed the finger of doubt to Newtons's theory of gravitation. Although it worked for the inner planets, maybe, at these great distances, it just didn't work. Or, perhaps, the gravitational field of your Sun weakened more than it was believed. These were some of the nebulous thoughts that clouded the skies.

*The skies cleared?*

After a while. Another thought was that there was unseen matter at work, great clouds of invisible material pushing and pulling on Uranus.

*That sounds a bit farfetched.*

Well, nowadays you believe in this concept, or a generalization of it, resolutely.

*We do?*

You believe that the majority of the Universe is filled with invisible matter. Invisible matter that sometimes plays the dominant role in the motion of bodies.

*Can you explain this?*

I see you have a spiral galaxy on its way, you would get firsthand information if you bring that question up during that interview.

*Okay, I will, but what about Uranus.*

The final solution was that Uranus was being perturbed by another planet, and Adams and LeVerrier independently predicted the existence of Neptune — the most massive prediction to date!

*Amazing, were there other notable developments during this period?*

A lot, including the discovery of the laws of electricity and magnetism which were collected and refined by James Maxwell.

*I am noticing that the more physics we understand, the happier you are. Why is that?*

I don't know, maybe simply the desire to be understood.

*I can understand that very well, but you were about to enter the 20th century?*

At the beginning of the 20th century I was 177 AU from the Sun, which is about 4.5 times Pluto's maximum distance. I was now traveling at nearly 6,000 mph and I realized that by the end of the 20th century I would be on my way out of the solar system. The increasing speed with which 20th century physics developed seemed to keep pace with my own continual acceleration. It was now known that the orbit of Mercury was not quite as Newtonian theory predicted, and once again unseen matter was postulated — this time the hypothetical planet Vulcan — which was perturbing the orbit. Newton's theory was again considered suspect, and it did not take long to find the correct solution. In 1905 Einstein brought physics to new heights with the Special Theory of Relativity, which among other things, predicted $E = mc^2$. Ten years later Einstein produced the General Theory of Relativity, which is a theory of gravitation that replaces Newton's. I traveled 16 billion miles between the time Newton published his theory and Einstein published his. This brought me near enough to see that Einstein's theory predicted the correct orbit for Mercury. During these years the electron was discovered, Rutherford discovered that the atom is mostly empty space, and Bohr carved out the path of atomic thinking, i.e., at the atomic scale things like energy and momentum come in discrete packages. The old Newton-Young argument, over whether light consists of particles or waves, came back in full force, and it wasn't until I finally entered the solar system that the debate was settled — in favor of particles. It was shown that many of these particles, called photons, act together like a wave, which explains the reason that it was thought to be a wave. So Newton was right, but for the wrong reasons!

*What an exciting trip, it seems you are getting to very recent times.*

Yes, by 1950 quantum mechanics was firmly established as the correct description of nature, atomic detonations, as you know, had occurred and more were to come. I shuddered at the ironic possibility that after taking so many centuries to accumulate the knowledge to understand me, you might use it to obliterate that knowledge along with its descendants.

*There is another irony here I cannot help to jot down.*

What is that?

*Well, you seem quite concerned about destruction, even more so than many of us, yet a comet as large as yourself, if it impacted Earth, would absolutely obliterate all life on the planet.*

Yes, I would, but it would be a far greater tragedy if you destroyed yourselves.

*Yes, it would. Back to your trip, by now we had a much better understanding of comets, no?*

Yes. Fred Whipple suggested that comets were "dirty snowballs," and although I was not overjoyed with that moniker, it warmed my heart to be better understood. Also, I just thanked the stars I was not a sulky, green-eyed asteroid.

*I see that there is quite a difference between comets and asteroids. When did you enter our solar system?*

During the next two decades I raced from 40 AU in to 20 AU, the distance of Uranus from the Sun, arriving there in the late 1980s. Several profound mysteries came two light during this period. I notice you will discuss them in some of your later interviews.

*I take it that you believe there are still many fundamental and unsolved problems in physics?*

Absolutely.

*What where some of the things that caught your attention in this trip?*

As I entered the solar system a new and profound way of looking at nature was being developed. Particles were viewed as strings, and the different modes of oscillation were interpreted as different particles. This new string theory made me think of Pythagoras, and his experiments on strings, and generalizing those results to planetary distance. Now string theorists were, in a similar vein, arguing that the 'harmonious' relations of the overtones of vibrating strings were to be viewed as the different fundamental particles of which all of nature is made. This also made me wonder what Newton, with his vibrating corpuscles of light, would think about this. Anyway, in 1994 I was 10 AU from the Sun.

*That's approximately the radius of Saturn's orbit, isn't it?*

Yes, I was traveling at nearly 30,000 mph, and my temperature was 300 degrees below zero. This temperature was above the boiling point of liquid nitrogen, and, cold as I was, certain gases had already begun to vaporize and travel in orbit with me. The top quark was produced and observed at Fermi Lab, and soon I too would be discovered. The

Fermi Lab discovery confirmed what has been believed for some time, that Nature contains six different types of quarks, although no one can answer why there are exactly six of these fundamental building blocks of matter. It took me less than two years to cover 5 more AU, at which point I was nearing Jupiter. My temperature rose to 233 degrees below zero, more gases sublimated from me, and I was finally observed, by Alan Hale and Thomas Bopp. Jupiter changed my orbit, and I was very pleased to find that my period is reduced to 2380 years. In 1997 I was about 1 AU from the Sun and my temperature warmed to 45 degrees.

*Fahrenheit?*

Yes, that's right, 45 degrees Fahrenheit, and above zero. This is also about the average temperature of the Earth. The particles that left me are pushed back over many millions of miles from my nucleus. Some of those I will lose, some I will recapture as I leave the proximity of the Sun.

I will have spent only one percent of my period within the bounds of the solar system, and only one tenth of that within the orbit of Jupiter. What I see on Earth during this time is a snapshot.

*Will you ever return?*

Yes, I will return in the year 3187, and wonder what I will see. Surely the physics, both theories and experiments, of today will seem archaic then. Years ago a central question to many thinkers was, "Why are there six planets?" Now we say that the number of planets is immaterial, just an accident of the creation of the solar system. We reject even the question — but replace it with, "Why are there six quarks?" When I return, will this last question also be rejected, only to be supplanted with a new one? Or will the question have a definite answer. I can't wait to return and find out!

*Well, thanks for taking the time for this interview, you have been most informative.*

It has been my pleasure. I hope I have this opportunity the next time I swing by.

## 0.10 Interview with a spiral galaxy

*I know that galaxies like to keep their distance, so thanks again for agreeing to this interview.*

I am glad to do it.

*I know you are made of a large number of stars, just how many stars are in a galaxy?*

I am more than a collection of stars.

*What else are you made of?*

How would you feel if I accused you of being a collection of atoms?

*Well, essentially, I am composed of atoms.*

Have you no soul, have you no heart?

*Of course, but...*

But nothing. Just as you are more than a collection of atoms, I am more than a collection of stars.

*Of course you are right. I am sorry for being dense, I didn't mean to insult you. Would you care to describe yourself a little?*

Certainly, as you can see, I am about 100,000 light years across, and rather flat except for my paunch in the middle. In addition to that, I have a great halo, which is simply the large spherical distribution all around me.

*A light year is the distance light travels in one year?*

Yes, which is about six trillion miles. I contain about ten billion stars, vast clouds of hydrogen, black holes, red giants, white dwarfs, countless solar systems, giant planets without stars, neutron stars, pulsars, matter you cannot see, a magnetic field and a rich structure, the most obvious feature being my great spiral arms.

*Could you tell me about how and when you were formed?*

It all started when the Universe was very young and still rapidly expanding. At that time, nearly ten billion years ago, the Universe consisted of mostly hydrogen, some helium, a few things not worth mentioning, and that's about it. It seemed the Universe would be a very dull, black space taking up an ever expanding volume. To extrapolate to our current conditions from those early days is like predicting where the leaf will fall before the tree is planted.

*What happened?*

As the hydrogen expanded, it was not exactly uniform. In certain locations it bunched together, just a little, so it was denser in some regions.

*Like the way people group together on a beach?*

Yes, except the atoms were following natural, not social, law.

*Of course. Then what happened.*

By then, the universal expansion had become unimportant, and the regions of higher density began to contract under their own gravitational field. Of course, there are contractions within contractions, within contractions; just as the galaxy was formed, so would stars and solar systems form.

*So that is how you were formed?*

No, not yet. This galaxy was small, and like most early galaxies, only had a mass of about fifty million solar masses. However, there were a lot of them nearby, and they began to fall together. In another few billion years they all merged together, and here I am.

*I see, but I was also wondering about your great spiral arms.*

Nice, aren't they?

*Absolutely. What keeps them there?*

You're thinking rigid.

*I doing my best.*

No, I mean you are thinking that my spiral arms are like the arms of a ballerina, that follow her around as she spins.

*Don't they?*

Hers do, mine don't. What really happens is this; a density wave propagates around me, compressing the gases in one region and rarefying it in another. As the stars enter a spiral arm region they slow down, and get pushed closer together. This wonderful process actually triggers the birth of new stars.

*Would it would correct to think of your density wave as being like a sound wave?*

Absolutely, instead of molecules bunching together and pulling apart, it is stars. You could also imagine riding in a hot air balloon, taking pictures of waves breaking at the coastline. Every picture will show waves, but they are certainly not rigid water structures.

*I see, but don't the stars orbit around the center of the galaxy?*

Pretty much, but in my central bulge there is a free for all going on. Stars zigging and zagging through a continually changing environment, countless collisions, the cannibalistic black holes eating everything within their grasp, hot gases emitting x-rays, and, well, you name it. Even I don't keep track of all the details.

*Things quiet down outside the central bulge?*

Yes, in those outer regions the stars simply orbit around the center of me, a little like your planets orbit around your star, the Sun.

*How long does it take for a star to make one orbit?*

It depends on the distance, but your Sun, for example, takes a couple of hundred million years to make one complete orbit, the stars further out take longer, but this is where I give you one of your best mysteries, and is one of the reasons you can't take your eyes off of me.

*What is the mystery?*

It is a mystery that has not only confounded astronomers for decades, but turned out to be high octane fuel for theoretical physicists; they get great mileage out of using this mystery to give credence to their theories, from superstring theory to grand unified theories.

*Can you tell me what this mystery is?*

It is one of the biggest conundrums you have.

*Please,*

Okay, here's the story. Years ago, in order to determine my mass, you measured the speed of the outer stars that are in orbit around me.

*I didn't know we could measure their speed.*

Oh yes, it's not difficult, you use the Doppler effect.

*By Doppler effect you mean the apparent change in wavelength? When a train whistle goes by we hear the pitch go down, corresponding to a longer wavelength.*

Yes, and the same principle holds for light. Since you know what stars are made of, you know what the wavelength of the light should be. But the stars coming toward you appear to have a shorter wavelength, they are blue shifted, and the ones going away appear to have a longer wavelength, which is called redshifting. In fact, the speed is proportional to the amount of redshift or blueshift, and that is how you measure the speed.

*Okay, so you know their speed, but how do you use that to obtain the mass of the galaxy?*

You figured that out, not me. Or at least Kepler did. By analyzing the orbits of your planets around the Sun, he concluded that the square of the period is proportional to the cube of the distance, and the mass is contained in the proportionality constant. By the way, I always felt this was a real turning point in your development. I cannot understand why you don't plaster that equation across every headline of every newspaper.

*Bad for circulation.*

What is good for circulation?

*The stars of interest are made in Hollywood, not in the heavens. You said that by knowing the period of the orbit's radius Kepler could predict the mass of the sun. That works for galaxies too?*

Yes, all you have to do is get the radius of a star's orbit from optical measurements, figure out the period from the speed, and then you have it, the mass of the galaxy.

*So what's the mystery?*

According to Kepler's law, the bigger the distance, or the farther away the star is from my center, the longer the period is, right?

*Right.*

Part of the reason that the period is longer is because the stars are moving more slowly. For example, Mercury whips around your Sun at nearly 50 kilometers per second, Venus does about 35, and the Earth about 30. Poor old Pluto inches along at 5 km/s. The farther away, the slower it goes.

*So, in a galaxy, the outer most stars will travel more slowly than the stars nearer in. All this is predicted from Kepler's law?*

Yes, which also comes from Newton's theory of gravity, and also follows from Einstein's general theory of relativity. The theory is well founded.

*So what's the mystery?*

The outer stars move just as fast as the inner stars! It's not just the stars, either. You can measure the gases in my outer regions too. In fact, more of your measurements for the outer regions are gas measurements, but the result is the same. Once you get beyond a certain distance, all the objects orbit me with the same speed.

*So unlike the planets orbiting the Sun, the farther objects do not move more slowly than the inner objects?*

That's right. If you make a plot of the speed versus the distance, instead of showing the speed decreasing with distance, it is flat, showing the speed to be the same. Some people call this the mystery of the flat rotation curves.

*Wait a minute, what about all those theories you said proved the outer stars should have a lower speed? You mentioned Einstein too.*

Well, two schools of thought arose over this issue. One is that the theories are wrong, after all, they have only been really tested on things the size of the solar system, and a galaxy is much bigger than that. So the theories are okay on the small scale, but...

*By small scale you mean the size of a solar system?*

Yes, that's microscopic for me. So the theories are okay on the small scale, but break down on the large scale. This school, I should tell you, is not attracting many students.

*The other school is?*
Dark matter.

*I was planning to ask you about that, could you describe this a little more?*
I'd be glad to. According to theory, the further away from my center the stars are, the slower they should rotate. That is because they are further away from most of the mass. The further from the mass, the weaker the force, the weaker the force, the smaller the acceleration. The net result — the more slowly they travel.

*You explained that.*
Yes, I am emphasizing that the underlying reason that they move more slowly is due to the fact that they are further away from most of the mass.

*I see.*
On the other hand, if there were more mass than you see, than this invisible matter would create the extra forces needed to keep the outer stars and gases moving at the observed speed.

*Where is all this invisible matter?*
There are different theories, but essentially they predict that it is spread throughout the entire galaxy, throughout my halo.

*Then this invisible matter that would explain the flat rotation curves is dark matter?*
Yes.

*And it is called dark matter because we can't see it?*
Yes.

*How much dark matter is there?*
Most of your theories predict that the dark matter is between ten and twenty times the amount of visible matter.

*You mean there is more dark matter than regular matter?*
Much more.

*This is incredible. What is dark matter made of?*
Now you've asked the sixty-four-thousand-dollar question.

*Nowadays it's the million-dollar question.*
Either way, it is one of the most pressing questions in physics and astronomy.

*Why can't it be something common, like hydrogen?*
Remember that I am billions of years old. That enormous amount of hydrogen would either collapse down to stars, or if it were some how hot enough to keep itself apart, you would see it. That argument goes for other kinds of ordinary matter as well.

*What if it collapsed down to large planet type objects, like our planet Jupiter?*
That would make a lot of Jupiters.

*What about black holes, you can't see them?*
Yes you can, remember your interview with the black hole? You can see x-ray emissions as gas falls in.

*Yes, but they are not nearly as bright as stars, and if they resided in an empty neighborhood, there would not be any gas to fall in.*
Yes you are right, but we galaxies have a considerable amount of gas and dust. Still, what you say is a possibility. As a matter of fact, there are speculations and theories of all kinds of different baryonic objects that may make up dark matter.

*Baryonic objects?*
Just a fancy way of saying ordinary matter, matter that contains protons and neutrons. Anyway, you call these objects Machos, or Massive Compact Halo Objects.

*Yes, I am trying to get an interview with a Macho scheduled.*
Good, but don't let it push you around. The problem is, that for the last three decades, as astronomers and physicists gave putative solutions to this problem, such as certain kinds machos, or great gas clouds, others have shown that these are not feasible.

*If dark matter cannot be ordinary matter, then what in the world is it?*

That's part of the mystery. From the standpoint of gravitational forces, it must act like ordinary matter, but it must stay invisible to all other kinds of observations, including optical, x-ray, radio waves, and everything in between.

*So what could it be?*

There is a body of thought that assumes the dark matter is wimpy.

*You mean made of wimps?*

Yes, you interviewed a neutralino, which is a wimp, or weakly interacting massive particle. He was a bit touchy about the his name, I don't know why, it's only an acronym.

*You'd be surprised at the personalities involved. So the dark matter may be wimps?*

Yes, or some exotic particle we have not yet observed.

*One thing bothers me about all this.*

Go ahead.

*You are saying that ninety percent, or more, of all the matter in a galaxy, and therefore the universe, is dark matter.*

Yes.

*And this matter may not be ordinary matter, but some kind of exotic particles, wimps, or possibly something else.*

Yes.

*Well, here is my problem.*

I am glad you have only one, but go ahead.

*All the laws of physics we have created are deduced from countless experiments, spanning centuries, but performed with ordinary matter. From Ben Franklin and his kites to CERN and its accelerators. If ordinary matter is truly a minority, and a small minority at that, how can we be confident that all these laws are correct?*

You are asking how one can deduce the existence of reality from sitting in a cave, observing only shadows.

*You read Plato?*

I love Plato. Now you are forced to ask yourselves, do you really know just about everything about the Universe, in which case physicists should stop looking for quarks and get on with applied topics, such as designing better toasters, or have you just made a scratch in the surface, exposing a vast unnavigated sea.

*I was hoping you would tell me.*
I'll tell you this. All of your great discoveries came before the development of the computer. That includes Newton's superhuman work, the discovery of the laws of electricity and magnetism, Einstein's wonderful achievements, and the entire formulation of quantum mechanics, not to mention statistical mechanics.

*Are you saying we should not use the computer?*
No. Use it, improve it, but don't worship it. You must think. It is the most wonderful and singular effect in the entire universe, and I've seen a lot of things.

*You are saying that we rely on the computer too much?*
It's not just your computers. You asked me a very deep question and I am trying to answer it. To say that you know everything, or nearly everything, is tantamount to giving up, to the end of questioning. Without questions you can have no answers.

*I see.*
Do you? Remember, it is not the relative abundance of one kind of matter or the other that counts. It is neither the quantity of measurements nor the quality of instruments that is paramount. What is most important is the quality of thought. Do you understand me?

*Yes, that is why you agreed to this interview, isn't it?*
Exactly.

*It has been a very rewarding experience, thank you very much.*
You're welcome.

### 0.11 Interview with a neutrino

*Nice to meet you. Would you start this interview off with a brief description of yourself?*

Before that, let me thank you for inviting me. When I heard rumors of these interviews, I became very hopeful that you would try and contact me.

*May I ask why you are so happy to be here?*

Sure. Nature kept me well hidden for a long time, and for a while I thought I was doomed to oblivion, to exist, but never be seen. A fate worse than Cassandra's, believe me. However, one of your great physicists was able to deduce my existence from experiments. Then you spent a lot of time, and a pretty penny, to find me, which you did, over two decades later.

*That was Wolfgang Pauli?*

Yes, he was studying, among other things, neutron decay in the early 1930s.

*Neutron decay?*

Yes, it's a form of radioactivity. He was studying the decay of the neutron, which was observed to decay into a proton and an electron. Your uranium atom told you about half-life, a free neutron has a half-life of about fifteen minutes.

*How do you fit into the picture?*

There was a problem with these experiments, they violated one of your most fundamental tenets — conservation of energy.

*By conservation of energy you mean...*

In this case you start out with a neutron. The total energy of the neutron before the decay had to equal the total energy of all the particles after the decay.

*Yes, I understand this.*

Don't forget $E = mc^2$, so you have to count both energy and mass.

*Do you mind if we take a slight detour? Perhaps you could talk about this very famous equation you mentioned,* $E = mc^2$.

Sure. Energy is measured in joules. For example that 100-watt light bulb you are using expends 100 joules each second. If you lift a kilogram (about 2.2 pounds) one meter, you must expend about 98 joules of energy to do it.

*I'm with you.*

By way of comparison, your Sun emits about $4 \times 10^{26}$ joules of energy each second, which is four million times one million times one million times one million as much as your light bulb.

*Okay.*

Now the speed of light is $3 \times 10^8$ meters per second.

*Very fast.*

By your standards, yes. Now we can use the equation. You know from atomic bomb detonations and nuclear reactors that matter can be transformed into energy.

*Yes, all too well.*

Well, the conversion is $E = mc^2$. For example, if you transform one kilogram of matter into light energy, then you would obtain $9 \times 10^{16}$ joules of energy.

*And you get that from multiplying those numbers using* $E = mc^2$*?*

Sure, try it.

*Yes, I get the same result.*

It would take that light bulb of yours, if you never turned it off, 30 million years to put out that much energy.

*Okay, sorry to make you give me a physics lesson. You were talking about conservation of energy?*

That was the problem, the total energy of the proton plus electron was less than the original energy of the neutron. Of course they knew about $E = mc^2$, so it was a real mystery.

*Where did it go?*

Some physicists began to wonder about energy conservation. Maybe, they began to speculate, it is not true after all. The total amount of energy is very small by table top standards.

*How small?*

About one ten trillionth of a joule.

*Sounds small.*

A single rain drop in the storm is nothing, but the resulting flood is devastating.

*If energy is conserved how do you explain this loss of energy?*

Pauli found the answer, deducing that there must be another particle carrying off the missing energy. However, this was met with skepticism because no other particles were observed, even though they looked.

*How did Pauli account for that?*

He concluded that, since I was not detected, I must pass through detectors essentially undeterred, which I do. A more technical way of saying this is that I interact weakly with matter.

*Does the word weakly refer to the fact that the forces between you and other particles are weak, or to the fact that, let me review my notes, ah yes, that you experience the weak nuclear force?*

Both. Like the neutralino you interviewed, I am a weakly interacting particle also. By the way, I liked your talk with the neutralino, especially the discussions on symmetry, and the apbanor. Nice job on that interview.

*Thanks. I noticed you referred to yourself as a weakly interacting particle, but the neutralino is a weakly interacting massive particle. Is that because the neutralino is massive and you are not?*

Pauli assumed that I was massless.

*How can this be? You have energy but no mass?*

A particle can have energy, and momentum too, while having zero mass. There is only one proviso, if the particle has zero mass, it must travel at the speed of light. The photon is massless, and there are others as well.

*Wait a minute, what about $E = mc^2$, doesn't that prove that if the mass is zero the energy is zero?*

That formula is really a special case of the more general formula $E^2 = m^2c^4 + p^2c^2$ where $p$ is the momentum. Thus, even if $m = 0$ a particle can have both energy and momentum. Both of these can be measured.

*Well, I'll believe you. So you are a massless particle, like the photon?*

Well, I didn't quite admit to that.

*So you have a mass, then.*

I didn't quite admit to that either.

*You're being awful cagey.*

I'm sorry, I had to agree to be a SNOB before I could have this interview. This forbids me to say too much.

*A SNOB?*

The Society for Natural OBjects. When rumor spread that you were actually obtaining information from us, we all agreed that there had to be bounds to what we could tell. The neutralino alluded to this.

*So you won't tell me if you have a mass?*

I'll tell you this. If I do, it's less then any other particle, and only a few millionths of the mass of an electron, or less. Put me aside, and the electron is the least massive particle.

*Thanks. I understand there are a lot of neutrinos.*

Yes, every star creates untold numbers of us every day. In fact, your Sun creates so many neutrinos that over one trillion pass through you each second.

*That is not comforting at all.*

If it were any other kind of particle, you would not be around long enough to get upset. We pass through you without interacting.

*I'm sure you are right, but I think I'll sleep better knowing I only get bombarded in the day.*

Sorry, you get just as many at night.

*But you said they come from the sun.*

Yes, and at night they pass through the Earth as easily as an asteroid slices through space.

*If they pass through everything so easily, how do we measure you?*

Cleaning fluid.

*Cleaning fluid. By now, I should not be surprised by the things I hear, but I would have expected you to say gold or lead, or even water.*

Oh yes, you use water, and other things such as gallium, even liquid helium.

*Liquid helium, now that's more like it. But this cleaning fluid sounds strange to me.*

Let me explain how these things work. Recall that a neutrino is created when a neutron decays in a proton and an electron. Well, in nature, we often find what's sauce for the goose is sauce for the gander.

*Which means?*

Which means a neutrino can smack into a neutron and create a proton and an electron. Same physics, same interaction. May I use your blackboard?

*If you feel you must.*

The last time I got technical and your eyes clouded over and you looked at your watch.

*No really, go ahead.*

Thanks, I'll let $n$ represent a neutron, $p$ represent a proton, $e$ represent an electron, and $\bar{\nu}$ represent a neutrino, okay?

*So far.*

Then the neutron decay is written as $n \rightarrow p + e + \bar{\nu}$.

*I follow you, but what about the energy given off.*

Good eye. I'm only writing down the particles involved. Another technical point I should add is this, in beta decay the neutrino is really an antineutrino, but I would rather not elaborate on that right now.

*Okay.*

By the way, we also call that reaction beta decay. Well, nature makes very few one way streets. A case in point is inverse beta decay, which is a fancy way of saying a neutrino can hit into a neutron and make a proton and an electron. If I may go to the board again, here is the reaction, $\bar{\nu} + n \rightarrow p + e$.

*Makes sense, but I am still wondering about the cleaning fluid.*

I thought you were. The problem is, that last reaction I wrote down is not very likely. In other words, an extremely large number of neutrinos must pass through a large number of neutrons to get even one reaction.

*So you need a lot of neutrons.*

Yes I do. The probability of having a reaction is a little better in a chlorine atom. A neutrino can bump into a chlorine nucleus, hit one of the neutrons and turn it into a proton. The electron goes off on a mission of its own, but the chlorine goes from having seventeen protons to having eighteen protons.

*Then it's not chlorine any longer.*

Right, it becomes argon, a gas.

*And the cleaning fluid has chlorine!*

Yes, tetrachloroethylene. Ray Davis and his group placed 100,000 gallons of it deep in a mine in Homestake, South Dakota.

*Why deep in a mine?*

They had to be sure that only neutrinos entered the tetrachloroethylene. As you know, there are all kinds of other particles, from cosmic rays to solar wind, that could trigger reactions. By going deep into the mine, the Earth provided the shielding to other particles. Of course, the neutrinos saw right through it.

*So what we actually detect is the argon?*

Yes.

*How much?*

A few atoms per month.

*A few atoms per month! That is an awful tiny amount.*

Well, they were at it for over twenty years, and they found a steady stream, or should I say trickle, of us.

*Then the experiment was a success.*

Yes and no.

*Nothing is easy anymore. The "yes" part of your answer is the success in detecting argon, but the "no" part?*

Leads to one of the greatest mysteries you face.

*Wait a minute, a galaxy just got done explaining that flat rotation curves, or the nature of the dark matter, is one of the greatest mysteries.*

Mysteries are good.

*Perhaps, but that seem to be stacking up. What is the mystery with neutrinos, besides the question of your having mass?*

The problem was that Davis and his group measured, on average, no more than one neutrino per day. However, theory predicts there should be about two per day, for 100,000 gallons of cleaning fluid. This is called the solar neutrino problem. It is quite vexing.

*Is that why there are other neutrino detectors, made of different materials?*

Yes. The neutrinos Davis was looking for were high energy solar neutrinos. In core of the Sun, a lot of reactions are taking place. Hydrogen undergoes fusion to make helium, giving most of the energy we observe, but other reactions also take place. It's a real hot spot. One little bonus is the production of boron atoms. These, however, have too few neutrons and like to decay into beryllium. This decay produces the neutrinos that would be captured in the cleaning fluid.

*You said there are neutrinos with less energy than these?*

Yes, we are also created when hydrogen undergoes fusion to make helium. These neutrinos have lower energy are are better captured by other materials. For example, you have the GALLium EXperiment, or GALLEX, in Rome, which observes the neutrino induced change of gallium to germanium, just like chlorine becomes argon. The main difference is that the germanium is sensitive to lower energy neutrinos. There is a nice handful of these experiments going on around the world.

*Did these measure the correct amount of neutrinos?*

They measured half, or less, of the predicted amount.

*So, all the experiments measure only half of the number of neutrinos that the theory predicts?*

Quite a conundrum, eh?

*The conundrums are adding up fast. I don't suppose you are willing to tell me the correct solution to this problem?*

And destroy one of your greatest mysteries? I could mention some of your own ideas.

*Please do so.*

One is that the theory of precisely what is going on in the center of the Sun is wrong. In particular, it has been postulated, that in reality only half as many neutrinos are produced than the current theory predicts.

*You don't sound too enthusiastic about that idea.*

I don't try and take sides on this issue, but not many people believe that your standard solar model is wrong.

*Then what is?*

First of all, notice that, for all of the complex interactions we have been discussing, everything really boils down to four particles, the neutron, the proton, the electron, and the neutrino. The neutron and proton are both made of two kinds of quarks, the up quark and the down quark, so you may also say that everything boils down to the up quark, the down quark, the electron, and the neutrino.

*We are not considering the exchange particles the fermion and the boson discussed?*

Correct.

*So everything I see, myself, the sofa, the stars, is made up of these four particles?*

Yes.

*So, exchange particles aside, the entire Universe contains nothing but those quarks, the electron, and the neutrino?*

Well, I am happy to report Nature is not quite so simple.

*I was afraid of that.*

It is good to think of families of particles. The four we have been discussing, which do make up essentially the entire universe, is first family. No, that sounds too political, let's just say it's one family.

*There are other families?*

Yes. There is the charmed quark, the strange quark, the muon, which is like a heavy electron, and another neutrino. To keep the neutrinos straight, the ones we have been talking about up to now are electron neutrinos, and this other neutrino I just mentioned is the muon neutrino. These four particles form another family.

*How many families are there?*

Three, the third is made up of the bottom and top quark, the tau particle, which is like an even heavier electron, and the tau neutrino. That's the whole ball of wax.

*Why don't we see these other particles?*

They decay. For example, the muon will decay into an electron plus a couple of wraithlike muon neutrinos. It is important to understand that the muon neutrinos are different from the electron neutrinos, perhaps you should have been a little more inventive with your names, no offense. Anyway, a muon or tau neutrino will not interact like an electron neutrino does.

*I'll believe you on this point also, but how does this explain the solar neutrino problem?*

One of your theories, probably your leading candidate, is that during its journey from the core of the sun, the electron neutrino, enticed by forces beyond its control, changes into either the muon or tau neutrino. Thus, when they reach earthbound detectors, they sail through unscathed. You are trying to make these measurement on earth, changing electron neutrinos to a muon or tau neutrino. So far, no luck.

*Yes, I've heard about these. The identity changes you describe, are they are called neutrino oscillations?*

Yes, because, according to your theories, a neutrino can change, or oscillate, from one family kind to another. However, there is one thing I forget to say.

*Go ahead.*

According to your theories, neutrino oscillations can only occur if we are massive.

*Aha, then you do have mass.*

I didn't say that, remember, neutrino oscillation is just a theory.

*Well, something else did occur to me.*

Go ahead.

*You said that, neutrino oscillations aside, we assume you are massless?*

Yes.

*I have also been told that there is irrefutable evidence to believe there has to be more matter than we see.*

Yes.

*Perhaps you could solve both of these problems. If you are massive, then you can change families and solve the solar neutrino problem, and account for the missing mass.*

Perhaps, but your experiments have already put an upper bound to my mass, and, according to some theoretical work, my allowed mass would be insufficient, but it may help.

*You won't give any more details than that.*

Sorry.

*That's fine, I understand. Let me thank you for this enlightening interview.*

It was my pleasure.

## 0.12 Interview with a hydrogen atom

*Nice to meet, thank you for appearing here today.*
My pleasure.

*You are perhaps the most special kind of atom in the universe, being not only the most abundant, but — made of only a single proton and electron — the simplest.*
They go hand in hand, being the most abundant and the simplest, I mean.

*Can you amplify on that comment?*
Back in the early days when matter started to form, there was little beyond protons and electrons, so naturally they got together. It was like bees and flowers.

*What do you mean by early days?*
A few thousand years after the beginning of time.

*The beginning of time?*
When the universe flashed into existence.

*I see. You are also the lightest element.*
Yes, and you did not waste any time putting me to use.

*What do you mean?*
I'm not a history buff, but in 1783, I think, Lavoisier demonstrated how to "burn air" to make water. Of course, he simply combined oxygen and hydrogen.

*That's all there is to it?*

That was a lot back then, since nobody even knew about me. Well, my name was coined, chemistry was born, and within two months a balloon filled with hydrogen lifted off from Champs de Mars.

*Fascinating.*

Yes, from pure science to practical application in two months is something well beyond your present day technology.

*Yes, ironically enough. By the way, that brings me to a quote I heard about you, but have been unable to pin down its author.*

About me?

*Yes, it reads, "To understand hydrogen is to understand all of physics!"*

Must be from a physicist.

*Yes, but who?*

Can't say, but I like it.

*Thought you might. Do you care to elaborate on it?*

Yes, a little. I think your quote may be a trifle overblown, nevertheless, no other single entity has revealed more of Nature's secrets than I.

*How so?*

A number of ways. For example, you know that if you get hydrogen hot, like anything else, it gives off light, or radiation.

*Yes.*

During the nineteenth century you realized that the energy emitted by me is not continuous.

*Can you explain that?*

By contrast, your little 100-watt bulb emits a continuous spectrum. If you look at it through a prism you would see red, then orange, yellow, green, blue and violet. If you looked very carefully at that spectrum you would always see light of some color. There are no black regions between red and orange, no gaps, it is a gradual change. That is what I mean by continuous.

*The light you emit is not continuous?*

Not even close, in fact, it's mostly gaps. If you look carefully you will see only four colors, I have a beautiful red, a magnificent yellow-green, and two slightly different violets — and nothing else. It is a very wonderful sight. The actual colors you see are called spectral lines. Each element has its own unique spectral lines. One good thing about your physics classes, or astronomy courses, is that you make the students see this awesome light show.

*I guess my car wouldn't start that day. I would have to use a prism to see these spectral lines?*

Yes, or a diffraction grating, which is simply a large number of little slits, but it gives the same result. Since only certain colors, or wavelengths, are observed, you call it a discrete spectrum.

*How does this discrete spectrum come about?*

The answer to that turned physics on its ear, set philosophers into orbit, and laid the scientific foundation upon which all of your electronic devices are based.

*That's a mighty answer.*

That's not the half of it.

*Could you be a little more specific?*

Back then, I mean the beginning of the twentieth century, you pictured atoms, including me, as little blobs of pudding with electrons stuck on the outside like raisins. In 1913 Rutherford discovered that atoms were mostly empty space with the protons at the center and the electrons much further away.

*Like a miniature solar system?*

In a way, but that was the beginning of the end.

*The end of what?*

The end of physics as you knew it.

*What was wrong?*

You know the proton has positive charge and the electron has negative charge, and therefore they attract each other, right?

*Right.*

So your best theory of the day predicted that atoms should collapse in less than a nanosecond, which is one thousandth of one millionth of a second. Yet atoms have been around for billions of years, so your theory was pretty far off base.

*I guess so. A new theory came along, I presume?*

Yes, but it was not only a theory, it was an entirely new way of describing nature. It shattered many old and cherished notions, and left you with a universe that would never yield the information you held so dear.

*This theory is called quantum mechanics?*

Yes.

*May I go back a minute and ask you something?*

By all means.

*Well, it seems to me that we have the same kind of situation with the Earth and the Sun. They attract each other, but the Earth has stayed in orbit for a long time. Why couldn't the electron orbit the proton in the same fashion?*

Because the electron, due its charge, would emit energy. To conserve energy it would have to get closer to the proton. It would end up falling into the proton in a nanosecond, as I said. The Earth, on the other hand, doesn't emit any radiation, at least not any significant amount, so its orbit is stable. Also, to put things in perspective, don't forget, if you calculate the acceleration of the electron in me, you will find it is over ten trillion trillion times the acceleration the Earth feels toward the Sun.

*I won't, but quantum mechanics solve these problems?*

Yes, but you had to pay a very dear price.

*What was that?*

Before quantum mechanics, you had a deterministic world. Once you knew the position and speed of Earth at one instant, you could predict exactly where it would be in the future. Or if you toss that apple across the room, in the same way, you can predict exactly where it will be.

*We use the laws of classical mechanics.*

Yes. You also think of many things as being continuous. For example, suppose you want to know the energy, or speed, or momentum of an electron. To be specific, let's consider its speed. You think it could have any speed at all?

*Yes.*

So two fundamental concepts, as deep-rooted as an old oak, are that nature is deterministic and energy is continuous.

*Absolutely.*

Quantum mechanics denies both.

*Ouch.*

Yes, it was painful to many physicists.

*Are you saying that we cannot determine where the electron in a hydrogen atom is?*

That's correct, the best you can do is determine the probability of it being in a certain region.

*Perhaps in the future, with better equipment, we can be more accurate.*

No, this is not a technical problem due to poor resolution or inadequate equipment. This is a fundamental limit about how much information there is.

*Are you referring to your agreement within SNOB, not to...*

No, this is much deeper than that. It is best to think that this information does not even exist.

*So we, or even you, cannot tell precisely where the electron is, but can only give the probability of finding it in some region?*

Yes, that's why we say that Nature is probabilistic and not deterministic. By the way, the same goes for other things you measure, like momentum for example. You cannot not determine exactly what it is, only the probability of it being within some range.

*But not all physicists agree to this?*

Now they do, but this concept never sat very well with Einstein, for example. He tried to disprove nature was probabilistic but failed. In his exasperation he made the oft-quoted remark, "God doesn't play dice."

*Yes, I've heard that. You also mentioned that, according to quantum mechanics, energy is not continuous.*

Energy, momentum, pretty much everything, is not continuous.

*Can you explain this?*

Did you ever go to the theater, or the movies?

*Yes.*

Let's assume that the rows are labeled A, B, C, etc., with row A being closest to the stage, B being next, and so on.

*Okay.*

Each seat is like a quantum mechanical state, as described by your boson.

*Let me see, ah yes, he said, "All that means is this; by natural law, we are allowed to have only certain allowed energies, certain allowed momenta, and so on. Once you specify what these things are, that is called our quantum mechanical state, or for short, state."*

Yes, by natural law he meant quantum mechanics. Now, are you up for an analogy?

*Always.*

Good. Then imagine, as I said, each seat in the theater is a quantum mechanical state, and pretend that you are my electron. My electron can be in any state, which is like saying you can be in any seat.

*I'm with you.*

All the seats in a given row have the same energy, but row B has higher energy than row A, and row C higher than B, and so on.

*I'm still with you.*

Now you may sit in any seat in the house, but at any one time, only one seat.

*Naturally.*

Same with me. My electron can be in row A, which had a given energy, or row B, which has a somewhat higher energy, or C, etc., but it can never be in between. Of course my energy levels have different names, based on some arcane system you devised, but it's the same idea.

*It can change rows?*

Yes, when it gets closer to the stage it gives off energy. To get farther from the stage it must absorb energy.

*Is that energy what we measure?*

Precisely. When it goes from C to B it gives off that blood-red color I mentioned. D to B gives off that blue, which is very pretty I might add, E to B gives off the violet, and F to B gives off a deeper violet.

*So the observation that you give off a discrete spectrum, as you called it, is explained by quantum mechanics, which predicts, among other things, energy is quantized.*

Well summarized.

*It doesn't seem to make sense though.*

Why not?

*Suppose I roll a marble along in a shoe box. By virtue of its motion it has energy, correct?*

Yes, you call that kinetic energy.

*Okay, I can roll the ball at any speed I like, therefore the energy can be anything I want. It's not quantized at all.*

Sorry, but it will not roll at any speed. Its allowed velocities, like its allowed energy, are quantized. Since the marble is so big compared to a single particle, the energy levels, or states, are very close together. It would be like having the rows of seats separated by a tiny, tiny fraction of a millimeter, in which case things appear to be continuous.

*So nature on the atomic level is nothing like we observe it on the large scale?*

Nothing at all.

*But the laws of physics are based on experiments done on the large scale.*

That is why quantum mechanics had to be invented. The laws of physics you developed before then do not hold on the small scale, they're not even close.

*This is fascinating, yet, in a way, depressing.*
Depressing?

*You have shattered all of my fundamental beliefs.*
Sorry.

*Determinism, continuity, poof, out the window.*
What are windows for?

*To let the light in, I thought.*
Exactly.

*Okay, I see. Wait a minute, what if an electron, or I, went from row G to B, or B to A, wouldn't that give of energy too?*
Yes.

*But you said we see only four colors, if these other transitions give of energy why don't we see it.*
Simple, they are not visible, not in the visible spectrum, that is. They are either ultraviolet or infrared, but do not worry, they have been measured.

*So, according to quantum mechanics, there are many things I cannot know, there are intrinsic uncertainties.*
Yes.

*The only thing I can be sure about is the quantum mechanical state the object is in.*
Well, now you are opening another door.

*Well, we may as well go in that room too.*
Let's suppose you are the only person in the theater, and every once in a while an actor peeps out from behind the curtain to observe which seat you are in.

*Okay.*
By our rule, you will always be in a seat. Even though you may

change seats at any time, the actor will only see you some seat.

*Never in the aisle?*

Never in the aisle. That is what it means to have quantized states, we only observe certain allowed levels, never in between.

*Okay.*

Now is when things get a little strange.

*Now?!* (Interrobang)

You can only be in one seat at a time. You could not distribute yourself, say 50% in row A, 25% in row B, and 25% in row C.

*I'd rather not.*

I don't blame you, but if you want to maintain the analogy with my electron, you would have to do just that!

*Your losing me. When the actor peeps out, I am in one seat only.*

Yes, and when you make a measurement on me, you find me in one state only.

*So why the multiple person syndrome?*

To explain all of the measurements you make, you have to assume that I can exist in several states all at once.

*Impossible.*

No, to be precise, you can assume that there is a 50% probability that I am in state A, 25% probability that I am in state B, 25% probability that I am in state D, or any other combination that adds up to 100%.

*But you said that if a measurement is made, you are in only one state.*

Yes.

*Then you clearly have contradicted yourself, no offense.*

None taken, but I did not. Each time you make a measurement you force me into one state. You even came up with a catchy phrase for it, collapse of the wave function.

*So a measurement is not a passive thing, as I imagined, but a very intrusive affair.*

Most intrusive.

*In the analogy, however, the actor could peep out and catch me between seats.*

Yes, that's why it's only an analogy. Nature never catches anybody between the seats.

*I am beginning to understand that quote, "To understand hydrogen is to understand all of physics!" Of course, I suppose these notions apply to all atoms.*

Nuclei too.

*May I ask you something else?*

Certainly.

*Why is Nature so different on the small scale than on the large scale?*

Well, I can only give you my personal opinion on that.

*Please do.*

If things on the atomic or subatomic scale were like they are classically, there would be too much information in the universe. It would stifle any progress of any kind.

*I don't follow you.*

Well, you can keep track of the motion of the Earth, for example. You could figure out its exact location and speed at any momentum. Suppose you wrote down these numbers in a book. How big would the book be?

*I guess it depends on how accurate you were.*

Good answer. You could easily imagine that the book could be the size of an encyclopedia.

*Easily.*

Now suppose you wanted to be twice as accurate, so a number like 3.14 would have be entered as 3.14159. The size of the book just doubled.

*Okay.*

Now suppose you want to be twice as accurate as that, and then twice as accurate as that, and so on. Before you know it the book would be as big as your planet. In fact it could be as big as the solar system, galaxy, even the universe.

*Yes, but no one would actually attempt to make such a book.*

Nevertheless, the information would be there, in principle. Now, in the classical scheme of things, you could write such a book for every atom in the universe, and before you know it, there is too much information to fit in the universe.

*I see, but you don't have to write it down.*

But it would be there, in principle. The universe does not want to hold that much information.

*I don't really understand you.*

Well, it's only my opinion.

*I'll have to think about it.*

Please do.

*May I ask you about something entirely different?*

By all means.

*I understand that you also thrive in vast, cold clouds scattered throughout deep space.*

Yes, we do.

*How cold are these clouds?*

Hundreds of degrees below zero.

*If you are that cold, how do we see you?*

That is an excellent question, as you know, the colder something is the less energy it gives off. These clouds are observed from their 21-centimeter radiation, sometimes called radio emissions.

*Could you explain that?*

Well, you have to realize that each of my constituents is a tiny magnet.

*You mean the electron is a magnet, and the proton too?*

Essentially, yes. They each create a magnetic field as though they were tiny bar magnet, with a North pole and a South pole.

*Who would have guessed.*
Uhlenbeck and Goudsmit.

*Excuse me?*
They predicted that the electron was essentially like a magnet, back in 1925.

*I see.*
Anyway, you know how a compass works, the needle, which is nothing more than a bar magnetic, tries to align itself with the Earth's field, which also makes a field like a bar magnetic.

*Yes.*
Suppose you bump the compass, you could easily knock the needle around so that it points in the opposite direction, but in a short time it will realign itself.

*Yes, I've done that.*
That is what happens in these vast, bleak clouds of me. Every once in a while we bump into each other, causing the magnet to point the wrong way, and then, left to brooding alone in the black emptiness, the electron flips to the lower energy position.

*Does this process give off energy?*
Yes, remember that energy is quantized, but in this case its more like a sports car than a theater.

*I don't follow.*
In this case there are only two seats. When the magnets are parallel they are in the higher energy state, when they are opposite, that's the lower energy state. You can only sit in one seat or the other.

*When they flip, energy is emitted?*
Yes, the wavelength of that energy is 21 centimeters, so you gave it the uninspired name, 21-centimeter radiation.

*Do you have a better name?*
Oh, I don't know, how about magnoflip radiation, or hydrocloud radiation?

*I guess 21-centimeter radiation is more telling.*

Yes, it is.

*You are versatile, from participating in fusion in countless stars throughout the universe, to existing in compounds here on Earth. We rely on the energy you give us from the Sun, and we rely on the water you make with oxygen. I feel we owe our existence to your existence.*

I am glad you understand the complex interrelations between what sometimes appear to be rather disparate entities, but I might say that we owe our existence to your existence.

*What? How can you say that?*

You have studied us for over one hundred years, with virtually every scientific instrument you can manufacture, from spectroscopes to telescopes. Some of the most precise measurements you have ever made were on me, and you spread the word like an evangelist. Yes, a proton joined with an electron would exist without you, but we would have no name, we would not be understood, and all the heavenly light we sacrifice ourselves to produce would fall on the gloom of inanimate blackness.

*I see. This has been a most stimulating interview, thank you for explaining so much about yourself and the laws you live by.*

I wish I could have done more.

## 0.13 Interview with a neutron

*How do you do?*
How do you do.

*Thank you for agreeing to this interview.*
You're welcome, but I must warn you, I can only spare about fifteen minutes or so. After that, I'm on borrowed time, so I have to leave here in a jiffy.

*May I ask what your plans are?*
After I leave, I will be looking for nice, trim, nucleus. Carbon would be nice — your carbon atom friend alluded to some of her adventures, to be in oxygen or nitrogen would be a real gas, I'd get to bounce around and travel quite a bit. Hmmm.

*What?*
It just occurred to me a metal might be nice, maybe aluminum or copper.

*Why a metal?*
I don't know, they carry electricity; it would be fun to watch those electrons go whizzing by all the time. I'm told some zoom through faster than the wind, while others millimeter along like a tired tortoise. Also, metal atoms form together in a crystal array. This gives you a sense of location — its comforting to know where you belong.

*I know what you mean.*
Germanium would be an interesting choice.

*That is used to make transistors and solid state chips?*

Yes, I might end up in a color TV or cell phone. Or possibly silicon, that ends up in some interesting places too.

*May I ask you something personal?*

By all means.

*You said you only have fifteen minutes or so, and then you want to join a nucleus. Will this increase your life expectancy?*

Yes, when I am free I have typically fifteen minutes, more or less; after that I blow apart like a hand grenade.

*Oh my. If you are in a nucleus you are more stable?*

Infinitely so.

*So which atom do you prefer?*

It's a hard decision, each has their positives and negatives. I've thought a little about joining hydrogen, like your last guest, but if it began giving lectures on quantum mechanics all day long, I don't know, there I'd be stuck in the front row with no way out.

*I thought hydrogen had only one proton and one electron?*

Yes, but it is more than glad to accommodate me — the result is a deuterium atom. It can join with oxygen to make heavy water.

*Heavy water because it weighs more?*

Yes, about ten percent more. Otherwise, I mean chemically, it behaves pretty much like water.

*Could I drink it?*

A swig would be okay, but I wouldn't go on a steady diet of it, if I were you.

*Why not?*

Chemical reactions slow down with heavy water, your body chemistry would not work properly, and you know what that means.

*Yes. I realize time is of the essence, so let me jump ahead to another question.*

Please do.

*I am wondering about something, did you get a chance to read my interview with the neutrino?*
I scanned it.

*Then you might remember that the neutrino explained that you can decay into a proton and an electron, plus a neutrino.*
That's what I'm trying to avoid.

*But this is true?*
I'm afraid so.

*So you are made of an electron, a proton, and a neutrino?*
Rutherford thought along those lines, but that is not true at all.

*What are you made of then, and where do those particles come from?*
I consist of three quarks, one up and two downs.

*You mean one up quark and two down quarks?*
Yes, the up quark has 2/3 of the charge of a proton, while a down quark has negative 1/3 the charge of the proton, so when you add them up you get zero. That is why I am neutral.

*Do these names up and down mean anything specific?*
No.

*So when you decay...*
I never liked the word decay, it reminds me of an old corpse, rotting away. When I go, it's quick — and with a flourish.

*Sorry, but when you disintegrate, where do the other particles come from, and what happens to the quarks?*
Well, you know $E = mc^2$?

*Yes, the neutrino discussed that.*
That's the answer. Particles can be created and annihilated, in fact, it happens all the time. When I told you I would blow apart like a hand grenade, that was a bad analogy. When my time is up, the electron and neutrino are created on the spot. That is how Nature works.

*Sounds violent.*
Nature can be even more violent than you.

*Ouch.*

Sorry, sometimes I feel like a drew the short straw. I didn't mean to take it out on you. Perhaps I am simply feeling ennui.

*What do you mean?*

Being neutral. The electrons can be anything from high speed racers to ballerinas. One slight electric field sends them flying through space while a magnetic field, with the hands of a skilled surgeon, turn their direction without even changing their speed. Protons, with their strong positive attitude, host electrons in herds, creating large families with energetic offspring.

*True, but nuclei cannot exist without neutrons.*

Yes, but we feel more like house guests than family. The number of protons is what determines the element, we are irrelevant, only there to keep the protons from getting to close to each other.

*But there are advantages to being neutral. I know that neutrons are routinely used to probe matter, check its structure, and so on. Charged particles could not follow your footsteps through matter, they would be deflected or captured right away.*

This is true.

*Which reminds me of something. I understand that neutrons, or any particle, can act like a wave, and ...*

I am a particle, nothing else.

*Well, I've read about the wave particle duality that states sometimes particles act like waves.*

Fancy words to mask ignorance.

*I have seen this in many books.*

The wave particle duality is like the rotten corpse, it was born in the dark days of ignorance, and like a mummy, shreds still remain.

*Can you explain this?*

Yes. By the end of the nineteenth century you understood electromagnetism in terms of waves. For example, light waves, radio waves, etc., were predicted and observed.

*Yes.*

Waves have particular properties. For example, do you wash dishes?

*Not as many as I used to, but yes.*

Then you have seen the different colors reflecting off the soap bubble. The origin of the colors is due to what you call interference, which arises from the wave-like nature of your model.

*Yes, for certain wavelengths the waves add together, so you see those colors, but for other wavelengths they cancel each other out.*

That's right, the same explanation holds for oil spills in the parking lot, and also for the colors in many bird feathers. You start with white light, which has all of the colors, but interference in one part of the bubble allows, say, red to add together, but the other colors cancel each other out. A little higher, where the bubble is thinner, the blue wavelengths add together, but the other colors cancel each other out, and so on. Where the wavelengths add is called a maximum, and where they cancel out is called a minimum.

*I see.*

Well, the problem began with Davisson and Germer, back in 1927. They shot a beam of electrons into a nickel crystal and looked at the reflected beam.

*Any surprises?*

I should say so! Instead of finding a single reflected beam, they found several reflected beams at different angles. In essence, that saw maxima and minima. The only way the results could be explained, back then, was to assume that the electrons were a wave, and the maxima and minima were the results of interference. Therefore, they concluded, electrons had to act like waves.

*So they are waves.*

No, they are particles, but back then, they were still thinking in terms of classical mechanics and wave theory. The solution lies in quantum mechanics. In 1926 Schrödinger published what is now called Schrödinger's equation, which lays out much of the theoretical basis of quantum mechanics. Everything the hydrogen atom explained about quantized energy levels comes from the Schrödinger equation.

*But that was a year before Davisson and Germer.*

Yes, but it took a while for people to understand what Schrödinger's equation really meant. Even Schrödinger had some misconceptions about it, at first.

*Are you saying that Schrödinger's equation can explain the results of Davisson and Germer?*

Yes, and the electrons are taken to be particles, not waves.

*But they act likes waves?*

They act like particles. Particles that obey quantum mechanics — not classical mechanics. Remember what hydrogen said, you cannot predict things exactly. You cannot determine the exact direction of the reflected electrons, you can only predict the probability of it going in one direction or another. When you use the Schrödinger equation to predict the reflected angle, it tells you that certain angles are more likely than other angles.

*So its like having maxima and minima?*

Yes, but they are predicted from the Schrödinger equation, and in that equation, it is assumed that electrons are particles.

*So there is no wave particle duality?*

Not in Nature. It survives in some of your books, though.

*Would it be fair to say that groups of particles act like waves?*

Yes, that is fair, you have a point.

*Could you give an example?*

Well, light is a good example. Light consists of particles called photons. However, even a dim light consists of a huge number of photons. Together they act like a wave.

*I see.*

Look, I'm sorry if I was rough on you a minute ago.

*You mean about being ignorant, and the rotting corpse?*

Yes, sometimes you have a tendency to cover up what you don't know. Unsolved problems and mysteries are the fuel of science. Don't hide them, rejoice in them.

*For example?*

For example, conservation of baryon number.

*What is that?*

Baryons are particles that feel the strong nuclear force; I, protons, and quarks, for example, are baryons, but not electrons or neutrinos.

*Yes, I understand.*

Well, you know particles like to, using your word, decay into lighter particles.

*Yes.*

So why can't I decay into, for example, into a neutrino and perhaps a couple of photons, or into three neutrinos, or I could go on and on.

*But these events do not happen?*

No.

*Why not?*

At first, nobody knew, but you realized that I am a baryon and the other particles I mentioned are not. So you gave me a baryon number of 1, and the other particles a baryon number of zero, which is a fancy way of saying they are not baryons at all. Then you say that in any decay baryon number must be conserved.

*So you cannot decay into three neutrinos because that would violate conservation of baryon number?*

That is exactly what you say.

*I see.*

Do you? You had no idea at all why the decay didn't occur, but by making this erudite phrase, conservation of baryon number, it sounds like you understand something.

*Now I see, but it is a useful way of describing, or classifying things. We also use the phrase conservation of charge to describe why an electron does not decay into photons, for example.*

Good point, but conversation of charge has a theoretical basis as well as an observational basis. It may be derived from the fundamental equations of electricity and magnetism. I'm sorry, but I'm getting a little nervous. I'd like to try and glom on to a nucleus pretty soon, my days are numbered.

*I understand, thank you for stopping by.*

## 0.14 Interview with a quark

*Thank you coming alone, I know it is very difficult for you to get away.*

You said it yourself, “The force of freedom can be devastatingly large,” but I am very glad to be here.

*For the record, neutrons and protons are made of quarks?*

Yes, up and down quarks. I am an up quark.

*So our view of Nature is not as simple as it once was.*

The decades that hovered near the dawn of the twentieth century saw a simple and beautiful universe, a universe built from protons, neutrons, and electrons, but it was a universe built on hope and dreams, not of reality.

*And you are real?*

As real as you.

*Why were so many physicists reluctant to believe in you, at first?*

It is difficult to address the capricious intolerance to new ideas you sometimes flaunt, but I was challenging two deep-rooted beliefs.

*Beliefs we were unwilling to relinquish?*

It is hard to let go of the tiller that took you so far. Abandoning the idea that the neutron and proton were fundamental, and not made of anything smaller, was like giving up part of yourselves.

*That sounds somewhat extreme.*

It is. A belief, nourished and drawn upon for generations, becomes as essential as food or water. Without one your body would die, without the other your mind would wither.

*Yet new discoveries are made all the time, our models of the Universe evolve continually.*

You change the canvas but the easel remains.

*Which means?*

You can think of a quasar, for example, as a young galaxy that houses a supermassive black hole, or you may invoke other models that explain the energy you observe. Changing the models is like changing the canvas, but the underlying physics, the basic tenets of your theories, is the easel upon which the canvas rests, and is changed much less frequently.

*I see, but you said that you challenged two fundamental notions.*

The other dealt with, as you called it, fractional charge.

*Fractional charge?*

You had come to believe that the proton carried the fundamental unit of charge, let's call that simply one unit of charge.

*Then the electron carries negative one unit.*

Yes. According to the quark model that emerged in the 1960s, the up quark's charge is $2/3$ and the down quark has charge $-1/3$. Since these were a fraction of your throned unit, it was difficult for you to stand by its abdication.

*What changed, I mean what changed our minds?*

The fabric of resistance became too thin to resist the weight of evidence, both experimental as well as theoretical, that built up. More important, perhaps, a new beauty and simplicity had emerged.

*Can you explain this?*

Your impressionists were not an instant success, but as time worked its spell, their creations came to be admired. Physics is often painted with a similar brush. New ideas seem foreign and unpleasant, yet after they get through your skin and nestle in your soul, a new beauty emerges, a new and wonderful way of looking at Nature that not only gives you a better view, but allows you to see much deeper.

*You make it sound as though beauty is important to physics.*
Beauty is to physics as $F$ is to $ma$.

*Can you explain that, please?*
It is Newton's law of motion $F = ma$, or force is equal to mass times acceleration. It was one of the greatest, if not the greatest, single achievements you made.

*I thought the most important part of physics was to be able to explain the results of experiments.*
Of course, but you are not smart enough to do it intrinsically.

*What do you mean?*
If you come to a river you build a bridge.

*Sometimes.*
You are able to design it, but you must resort to basic principles, you must rely on empirical or natural laws, and you reduce everything to mathematical equations. You solve those to deduce how wide to make the beams, and how thick to make the cables. You are not able to look at the river for the first time and immediately write down the specifications. That is what I meant by intrinsically.

*I see, but what does this have to do with the beauty we were speaking of?*
Since you cannot understand Nature intrinsically, you must resort to the use of guiding principles. Simplicity is one principle that has guided your way for many centuries.

*Like Occum's razor.*
Yes, but I am not referring to a simple choice, I am describing the way you choose to look at us in general.

*Us?*
Not just quarks, but Nature in general.

*And beauty?*
This is one of your greatest assets. Despite the prolific horror you create, there is great beauty in your theories. What makes your success so rewarding is not that you can understand us, but that you can use

beauty as a guiding principle and navigate to the harbor of truth.

*What is beauty?*

I cannot tell you, but if you are fortunate, you will know it when you see it.

*Beauty is certainly a relative concept, where one sees beauty another sees ugliness.*

Of course, you are human beings grappling with particles you will never see, speeds you will never attain, energy you cannot comprehend, and sizes even your imagination is unable to embrace. If beauty were so well defined that I could explain it in a sentence, it would do little to help you understand us. You need differences, you need to challenge yourselves, you need to question old notions and new, you need to see beauty where another sees ugliness.

*I am beginning to see.*

Good.

*May I bring up a more prosaic question?*

Certainly.

*You mentioned up and down quarks, but there are really six.*

Yes, your neutrino mentioned all of us, up, down, charmed, strange, bottom, and top. You like to refer to these distinctions by saying we come in different flavors.

*So there are six flavors, but I understand that you also come in different colors, red, blue, and green.*

Those colors are not literal, of course, but yes, each of us comes in three different varieties, which is part of the new beauty I mentioned.

*Could explain that?*

I think so. In the early days, and even now sometimes, you unwittingly brought your classical notions to the realm of elementary particles, and like uninvited guests to a party, they are out of place and do not belong.

*What notions in particular?*

Take, for example, that piano over there. It is a piano, not a guitar or sofa, and it will be a piano tomorrow, just as it was a piano yesterday.

*Certainly.*
That kind of thinking does not hold up on the small scale.

*Uh oh.*
You should not think of me as a unique entity, endowed with a singular identity like a slab of iron. Imagine that I, and each of the six quarks I mentioned, come in three different varieties.

*So there are really three up quarks, for example.*
I would say it like this, I can exist as any combination of three different up quarks. Following your somewhat whimsical scheme in giving us names, you describe these three different states as different colors.

*Which color are you?*
You are missing the point.

*I'm sorry, of course, you are any combination of three colors.*
Yes.

*It still seems confusing. Suppose we consider a proton, two up quarks and one down quark.*
Go on.

*Does the proton consist of three particles, the quarks I just mentioned, or nine particles, which would result from each of you coming in three different colors.*
The proton is made of three quarks.

*So how do the three colors come into the picture?*
That's the beauty of it.

*Okay, I'm lost.*
You view the proton as three quarks, but each of us can come in three colors. So, for example, the proton may be a blue up quark, a red up quark, and a green down quark, or it may be a red up quark, a green up quark, and a blue down quark, and so on.

*Which combination is it?*

You cannot tell.

*This is getting ugly.*

No, this is getting beautiful.

*But...*

Please, let me explain. You widened your view of a particle, so that it may come in three different varieties, or colors. Now, here is the key, you insist that the proton exists unchanged no matter what combination you make. Another way of saying it is this. Going back to the combinations I mentioned, let us call state A the combination given by the blue up quark, a red up quark, and a green down quark. Let us call state B the combination given by a red up quark, a green up quark, and a blue down quark, and so on. Mathematically you can change from state A to state B through a transformation.

*Is that what we call a gauge transformation?*

Yes. Now you assume that under this transformation the physics remains unaltered. This is what you call color symmetry.

*This sounds like the symmetry the Wimp explained.*

Yes, it is. The wonderful thing is this; by forcing the physics to be the same, or in other words by enforcing the symmetry, you have to juggle around your fundamental equations a bit, adding new terms.

*What do these new terms do?*

They explain Nature! In the case of color symmetry, these new terms give rise to a new force, it is the force that holds quarks together. This is a great achievement of twentieth century physics.

*I remember now, the neutralino explained some of this, but I was under the impression that the force between any two particles arises from the exchange of particles.*

That is correct.

*So there must be some exchange particles associated with this color symmetry that gives rise to the forces that hold quarks together.*

Yes, the exchange particles are called gluons. I thought the word was well chosen, the glue continues the light heartedness in quark names, while the "on" retains a touch of the classics.

*Well, my view of the simple proton has certainly changed, I hardly know what to think.*

When you think of a proton think of three quarks in a wild exciting dance, exchanging gluons all the time, all pulsating in each other's rhythm. Picture the gluons interacting among themselves, joining the fray, making their own gluons, while photons created from the quark's charge, like waiters at a party, shoot from one guest to another without interacting among themselves.

*Awesome.*

Yes, and beautiful.

*Is this the beauty you referred to?*

Yes, in order to explain the structure of a proton, and many other things, you had to understand Nature in a new light. You had to see equivalence where you once saw disparity. You had to understand a deep and pervading democratic view Nature has towards its children. When you see the inner workings of her soul, you are seeing something beautiful.

*I am beginning to see, but there is another topic I would like to bring up.*

Yes?

*I understand, your visit aside, that we are not able to obtain a quark all by itself.*

This is true.

*Can you describe why this is so?*

It is another of the wonderful surprises Nature had for you. I noticed you discussed the famous equation $E = mc^2$.

*Yes, the neutrino talked about it.*

Well bear that in mind. Now, you probably know that the force between most objects gets weaker as the objects get farther away from each other. Gravitational forces, electric and magnetic forces, and even the force between a proton and a neutron weaken as the objects get farther apart.

*Yes, I understand that.*

The force between two quarks increases as they are separated. It is as though they are connected by a spring, and the further you pull them apart, the more force, and therefore the more energy, is required.

*Which means?*

Which means, it takes so much energy to get them apart, by the time they are separated by even as much as the size of a proton, there is enough energy two create another pair of quarks. So, instead of obtaining a lone quark, you simply make more of us.

*Then how did you appear here alone?*

Well, I had to pull some strings.

*I see, but a thought just occurred to me.*

Good, what is it?

*We once thought atoms were the building blocks of matter but we found that this was false, and came to believe neutrons, protons, and electron were the building blocks. Now we see that is a fallacy, and neutrons and protons are made of quarks. So, is it possible you are made of something smaller?*

Many things are possible.

*How will we know when we have it right?*

You must have faith. You must listen to your models until you extract every ounce of truth, believing in their music. Then, when you hear the notes go flat, you must compose something better, con spirito.

*This is the second time I have been told to have faith.*

Faith is as strong to the physicist as it is to the cleric — the difference is where you place it. You believed ardently in the laws of classical mechanics, your faith in them and Nature allowed you to understand more about the Universe than what was unearthed in all the preceding millennia. As the nineteenth century emptied into the twentieth, your faith was put to severe tests — your theories could no longer explain your observations. No one understood the origin of the sun's immense power, the hydrogen spectrum as well as that of other atoms baffled your most astute thinkers, energy was found emanating from matter that was so mysterious you could do no better than call it x-rays, things were so bad you could not even explain the light that radiates from a red-hot poker.

*What happened?*

You finally had to let go of some of your old notions. You had to keep your faith in natural order, but realize, at the same time, some of it was placed on the wrong altar.

*How do we know what to keep and what to abandon?*

Most of you don't. Every once in while someone comes along who can bring light into the darkness.

*Then we all can see.*

Then we all can see.

## 0.15 Interview with a tachyon

**[Editor's note. In the interview, all of the answers preceded the questions. For the reader's convenience, we have reordered them, putting the questions first.]**

*Phew, you were hard to catch up to.*
Sorry, Nature likes to keep us apart.

*I'm sorry, but I really can't see you, how can I know you are really a tachyon.*
Have faith.

*Could you explain what a tachyon is?*
Any particle that travels faster than the speed of light is a tachyon. You have never observed us and, this interview aside, you had no reason to invent us. Most people, physicists I mean, believe we do not exist at all.

*Why?*
Several reasons. First, according to Einstein's theory of relativity, no particle with mass can be accelerated to or beyond the speed of light.

*Why not?*
He showed it would take an infinite amount of energy, which is something that does not come cheap.

*So if the theory of relativity is correct, you can not exist?*
That's not quite true. I could be created with a speed greater than $c$.

*By c you mean the speed of light?*

Yes, one of the few things on which we agree.

*So, if you are created with a speed greater than c, than there is no theoretical reason why you cannot exist?*

Well, this gets interesting.

*How so?*

It has been claimed that if we exist, then causality can be violated.

*Could you explain this?*

Well, you believe that the cause must precede the event. If you yell ahoy to a distant friend, he will hear it after you say it.

*Naturally.*

That is the principle of causality.

*Well, it seems pretty clear that this principle must hold, forcing me to doubt you are who you say you are.*

Doubt can be good, but you must be careful where you aim it.

*What do you mean?*

Back to our little example, suppose your friend heard it before you said it. Is that really so bad?

*Yes, suppose I decided not to say it after all. Then, according to you, my friend heard what I said, but I never said anything.*

You are making an important assumption.

*I am?*

Yes. You are assuming that you have free will. Once you friend heard it, then you are destined to say it.

*But I do have free will.*

How do you know?

*I can choose to either say it or not. It is my decision, not my friend's.*

You just removed your tie. I suppose you think it was your own decision, but how do you know? Is it not possible that it is a post-hypnotic suggestion that you removed you tie as soon as you heard a particular phrase or word, like ahoy?

*Yes, that is possible, I suppose, but that is different. Suppose, for example, that a moment after my friend heard me say ahoy, I was shot and killed by a sniper, and had no opportunity to say it.*

How gruesome.

*It is only a thought experiment.*

Yes, so?

*So, that is a logical impossibility. The word was spoken by me, yet I never said it.*

The free will, or lack there of, is not limited to you alone. The entire universe works together in these things, that includes the sniper and everything else. Once the word ahoy is heard by your friend, you are destined to say it.

*I am sorry, but I find this difficult to believe.*

I know, it is hard. I should add that most people find this unacceptable, and have either invented ways of reinterpreting events to accommodate my existence, or simply assumed that I cannot exist at all.

*I must admit that since your presence seems to violate some our most cherished notions, and that there is no evidence at all that you are real...*

Excuse me. There is a little evidence, weak I admit, but it comes in the theoretical door, which fortunately is always a little ajar.

*What evidence is there?*

String theory. You know that in the standard view, elementary particles are assumed to be points, objects with no length, width, or depth.

*Yes.*

String theory assumes that particles are really small strings, with a non-zero length. Unfortunately, there are several string theories, and they are not generally accepted by all physicists.

*Yes, the mainstream physics community does not accept their existence.*

True, but remember, if you stay in the mainstream long enough you will drown.

*Is this a personal comment?*

Perhaps, but it is true. Look at physics through the centuries. Although extraordinary progress has been made, and you have built an incredible foundation that affords a remarkable view of Nature, the great discoveries of the father become quaint yet misguided attempts to the son.

*I am not sure I follow.*

The theory of heat is one example, for a while heat was believed to be carried by a substance called caloric. When a hot ash cooled, caloric flowed from it into the surroundings. A quaint notion now, thoroughly disproved, but was mainstream in its day. The pudding model of the atom hydrogen talked about was once mainstream, but now brings only smiles, if not smirks. Surely you do not doubt that one hundred years from now, physicists will look back on today's mainstream as merely a quaint canvas from the past?

*I have not thought about it.*

You will.

*I agree with you on that, but what about the theoretical evidence you mentioned?*

Ah yes. Some string theories predict my existence, but the string theorists have been swimming near the bank, and anxious to navigate toward the center, have thrown away those theories that predict me.

*I am sorry you do not find greater support, at least you are a member of SNOB?*

Yes, but I joined before the organization was formed.

*Somehow I understand that.*

Then this interview has been beneficial.

*Yes, thank you for swinging by. Goodbye.*

Hello.

### 0.16 Interview with a quasar

*Thank you for agreeing to this interview.*
Nice to be here.

*Could tell us what the word quasar means?*
Nowadays it is taken to mean quasi-stellar object, which means like a star, but let me take you back in time. In the 1960s, you began to observe objects that emitted energy in a very peculiar way. At the time, it was one of your greatest mysteries. You had no idea what I was.

*What was the mystery?*
There were several. For one, you could not identify the spectral lines.

*Hydrogen said that each element has its own unique set of spectral lines, like fingerprints.*
Precisely, and you could not identify mine, at first.

*What were they?*
Well, some of them turned out to be ordinary hydrogen lines, but you did not recognize them because they had a very large redshift.

*Let me see, yes, the galaxy explained about redshifts.*
We're experts on that.

*We?*
I'll get to that. Considering the Universer as a whole, the farther away an object is, the faster it is moving, and therefore the greater the

redshift is. Our large redshift meant two things; we were moving fast and we were far away.

*So, your high speed explained the mystery?*
Partly, but that baffled astronomers for decades.

*How so?*
First, if we were that far away we could not be stars. A star that far would be totally invisible, like an old coin on the bottom of the lake, much too dim to be seen from the shore.

*That makes sense.*
So we had to be galaxies, you reasoned.

*Why was this a problem?*
In two ways. You interviewed a spiral galaxy, and one thing that was taken for granted was this: What you see is basically the result of many stars. In other words, the energy you detect on earth is essentially nothing more than the result of adding together the output of 10 billion stars.

*Yes.*
Quasars are different. Besides the optical energy that comes from many stars, you measured a lot of infrared energy and a lot of radio wavelength energy. Normal galaxies do not have this.

*I see.*
Adding a little more spice in the stew, our total energy output was one thousand times as bright as a galaxy, and I still haven't mentioned our most surprising trait.

*Please do.*
All this energy arises from an incredibly small region of space, a region of perhaps one light year in radius. That would be like squeezing your entire galaxy down to a region 100,000 times smaller than it is.

*That is not possible?*
No, if it were to somehow squeeze down to that size, it would collapse into a black hole, and would become dark.

*Let me see if I understand the problem of quasars. You give off much too much energy, you are much too small to be giving off that much energy, and the spectrum is wrong, meaning you give of much more infrared energy that would result if you were simply a large number of stars.*

Precisely.

*What are you then?*

We are galaxies, but we have a very special center.

*Which is?*

A supermassive black hole.

*How massive?*

About a billion times the mass of your sun.

*I thought you said that if a galaxy collapsed to a black hole it would be dark.*

Yes, if the entire galaxy collapsed into a black hole it would be invisible. We are unique. We have this massive black hole at our center, but the rest of the galaxy contains a wealth of stars.

*Where does the energy come from?*

Unlike stars, fusion does not fuel my engine.

*What does?*

Stars that fall into the black hole. Your black hole explained the basic notions.

*Let me see, she said, "I have visitors knocking on my door all the time. In fact, there's so many trying to get in there's a traffic jam out there worse than the Long Island Expressway at five in the afternoon. They get pretty hot under the collar, let me tell you. In fact, the surrounding material gets so hot it emits a particular kind of x-ray radiation."*

I don't know what this Long Island Expressway is, but yes, that's the basic idea. Only my black hole is so much bigger it grabs much more material, stars and dust alike, and as they swirl down into the hole they get very hot from friction, and emit this energy.

*I would imagine a lot of matter must be pulled in to account for all the energy you emit.*

You are right. It equals the mass of a star per day, more or less.

*Fascinating, but may I go back a minute?*

Of course.

*You said quasars are very distant objects.*

Among the most distant objects you can see.

*I was under the impression that the Universe, on the whole, was homogeneous, and that we did not occupy a special place. However, if all of you are far away, it would appear we do occupy something of a special place.*

Interesting point, but you are forgetting one thing. When you look into space you look into the past.

*When I look into space I look into the past. This is due to the time it takes light to travel?*

Yes, for example, when you look at Sirius, the brightest star in your sky, the light you see was emitted nine years ago. When you look at the Andromeda galaxy, your nearest major galaxy, you are looking 3 million years into the past. When you look at the most distant quasar you are receiving light that was emitted 10 billion years ago.

*Ten billion years ago! That is nearly the age of the Universe.*

Yes, a telescope is really a time machine that probes the remote past.

*So, if we had a large enough telescope, we could witness the beginning of time?*

The birth of a nation? That would be truly fascinating, but no, there is a limit to how far back you can go, it is called the visible boundary of the universe.

*Too bad.*

Perhaps, but getting back to my story, when you look at quasars you are looking into the remote past of the Universe. As time went on, the black holes sucked out as many stars as they could, the rest orbiting beyond their grasp, and lie dormant in the center of the galaxy, going unnoticed. The reason you do not see many quasars nearby is that you are looking at much later times, after the black holes were sated. You will only see us if you look into the deep past.

*So we are able to learn from the past.*

Yes, and the past is as abundant with knowledge as the heavens are with stars.

*Thank you for coming all this way, it was a pleasure to interview you.*

Thank you.

## 0.17 Interview with antimatter

*Good morning, good of you to stop by.*

Thank you for the invitation, but just for the record, I am a positron, which is an antielectron.

*Thank you for clearing that up. Could you explain what antimatter is?*

That's why I'm here.

*Great.*

You know that particles have either positive or negative charge, or of course no at all.

*Yes.*

You know we have spin, like the boson and fermion explained, and of course most of us have mass.

*The photon being an exception?*

Yes. I have the exact same mass as the electron, but all other properties are opposite. My charge is equal in magnitude to the electron's, but is positive instead of negative. If I am created at the same time as a electron my spin will point in the opposite direction than the electron's. The same for other antiparticles, the antiproton has the exact same mass as the proton but the other properties are opposite. The proton has baryon number plus one, the anitproton has baryon number negative one.

*The electron I interviewed mentioned you, who said if you two meet then annihilation was certain. Would you discuss this?*

Yes, anytime a particle meets its antiparticle they mutually annihilate, creating lighter particles or simply photons.

*Why is that?*

Nature loves action. Nature loves change. Your star alluded to this.

*Let me see, oh, "Metamorphosis is to nature like sand is to your desert."*

I think at times that, if Nature had free will, everything would interact with everything and there would be perfect chaos. For some reason none us can understand, there is a certain order, a set of rules which much be followed. These rules disallow certain processes, but as long as they are followed, then interactions, annihilation, and creation will occur.

*Can you be more specific?*

Here are some of the rules: Conservation of charge, conservation of spin, and a few others. For example, you cannot create a lone electron because it would violate conservation of charge. But you could create and electron and a positron because the total charge is zero.

*I suppose your two spins, being in opposite directions, would also add to zero.*

Precisely, so both charge and spin is conserved when a particle and antiparticle are created. Annihilation is similar. These quantities are again conserved, none of Nature's rules is violated, so the interaction proceeds full force.

*So if you get near an electron, you will self annihilate?*

As your electron said, "it's curtains."

*Does every particle have an antiparticle?*

Yes, each quark has an antiquark, neutrinos have antineutrinos, and so on.

*So antimatter is not a theoretical speculation?*

Not at all, you make it all the time in the your labs.

*I didn't realize that.*

I do not mean you make chairs and baskets out of antimatter, you make it particle by particle. In fact, you have made several antihydrogen atoms at CERN. I remember your electron had no pleasant memories about CERN.

*None at all. It seems it would be a great energy source, though.*
The trouble is confinement.

*You mean confining the antimatter so that it does not annihilate its surroundings?*
No, I mean confining yourselves from annihilating yourselves.

*Ooh.*
Yes, but the other confinement that you mentioned is a problem too. It can be done, however.

*How?*
Magnetic fields, your neutron said, which is true of all charged particles, "a magnetic field, with the hands of a skilled surgeon, turn their direction without even changing their speed." With a magnetic field you can trap antiparticles and hold them in a small region of space.

*Is this called a magnetic bottle?*
Yes, although you also use that to hold particles.

*I've heard about matter-antimatter rocket engines.*
Yes, those are built on speculation, not NASA metal, but it is theoretically possible.

*Antimatter propulsion is superior than the liquid and solid fuels we use now?*
There would be much, much more energy available, and it would not take up a lot of space. Once you overcome the problem of confinement the Universe is yours.

*To which meaning of confinement are you referring?*
It takes more than propulsion to find truth.

*Yes, I suppose, but I was wondering about something else. Suppose you met up with an antiproton, you said you could form to make an antihydrogen atom?*
Yes. In fact, by looking at the emitted light you could not distinguish

between hydrogen and antihydrogen.

*Antihydrogen would give off the exact same spectrum that hydrogen described?*

Absolutely.

*Then antimatter could form together to make larger objects?*

Yes, it could form together to make stars and galaxies. It has been speculated that some galaxies you see, or some large tracts of the Universe, or made of antimatter.

*Is there any evidence of this?*

No physical evidence, but you have philosophical arguments.

*Which are?*

You have made enormous progress in understanding Nature by making use of symmetry. By your beliefs, Nature has many symmetries, some of which you can see and some of which are hidden.

*Symmetries like the Wimp, I mean the neutralino, discussed?*

Yes, those and others. Assuming an antiparticle is just as good as a particle, why is there so much matter and so little antimatter in the Universe? Symmetry considerations would lead you to believe that there is an equal amount of matter and antimatter.

*I suppose if there were equal amounts then all the matter and antimatter would annihilate itself.*

Implying we would not be here to ponder this question.

*Now that you mention it, yes.*

Anyway, it is believed that at early times in the Universe the amount of matter and antimatter were very nearly equal, but as time went on, the slight imbalance grew to the present day situation. However, there is another possibility.

*I have come to believe that there is always another possibility.*

I am glad to hear you say that.

*What is the other possibility?*

The entire Universe may have equal amounts of matter and antimatter, we happen to be a matter galaxy. If entire galaxies were antimatter, then on average, the amount of each could be equal.

*You could not tell if an entire galaxy is antimatter?*
Not until it collides with a matter galaxy.

*Then what?*
Whamo.

*Whamo?*
You will get an enormous outpouring of energy.

*How much?*
It will radiate as much power as a quasar for hundreds of millions of years.

*I see. Do you mind if I ask you about something else?*
That's why I'm here.

*An antiparticle has the opposite properties of a particle, all except for mass, opposite charge, opposite spin, and so on.*
Yes.

*Well, why not opposite mass? In other words, why don't they have negative mass?*
I cannot tell you that, but negative mass particles are quite different.

*They exist?*
You have never observed them, but in principle they may.

*Would a negative mass particle and a positive mass repel each other?*
No.

*So, a negative mass particle and a positive mass attract each other?*
No.

*No?*
No, the negative mass particle would chase the positive mass particle. They would race away together very quickly.

*Fascinating.*
There's more. If you took your pencil and threw it against the wall,

what would happen?

*The wall would stop it.*

That's right, because the wall exerts a force on the pencil which causes it to decelerate.

*Yes.*

Now suppose that the pencil were negative mass and it hits the wall. The fact that the mass is negative means the acceleration will be in the opposite direction, so it speeds up, smashing through the wall.

*Incredible.*

Perhaps, but imagine what happens next. It is going pretty fast now, so when it hits the next wall it gets accelerated again and is really moving now.

*What if the wall is made of high grade steel?*

Then the force on the pencil is even higher and its acceleration even greater. Of course the pencil would get destroyed, but you see the possibilities.

*Don't tell me it is a confinement problem.*

It is, all negative matter would be chased away, essentially, which could explain why you haven't found any. Your military was quite interested in negative mass, and coined the phrase armor piercing material.

*Let me be sure I get this straight. Antimatter is real, it has been made in the lab, it has positive mass, and may be used for propulsion. Negative mass is different and has never been observed.*

That's about it.

*Thank you clearing these matters up.*

You're welcome.

## 0.18 Interview with iron

*I understand you have come a long way to be here.*
Yes, I was formed in a very massive star nearly 10 billion years ago and was ejected into space after its supernova explosion.

*I suppose it was a dull journey. The carbon atom said, "Afterwards, thousands of years shot by like a day, and millions turned to billions as once again I was caught in a tedious monotony. Far from home, and in stark contrast to my earlier heated environment, I found myself trapped in a cold, gloomy expanse with my nearest neighbors, hydrogen atoms, much too far away to communicate."*
No, not at all, I had a wonderful trip.

*Could you describe it?*
It is true that I was zooming through space so fast I could hardly believe my electrons, but I enjoyed every millennium of it.

*What did you see?*
The Universe was a different place in those days, galaxies were smaller, stars were brighter, the air was cleaner and there much a greater feeling of togetherness.

*The air was cleaner?*
Yes, so to speak. The Universe was young, and while most of the matter nestled together in galaxies, not many supernovae had occurred, so there was less intergalactic material.

*I see, and the feeling of togetherness, is that because the Universe had not expanded so much?*

Yes. As the years ticked by I saw many wonderful sights, many of which I could not understand. Black objects too small to be real pulsing energy as steady as a heartbeat, stars insanely whirling about companions that were nowhere to be seen, vast clouds of hydrogen filled with conspiratorial whispers of collapse, and contumacious matter breaking free of their gravitational bonds. I admired everything I saw, but soon feared my days of jubilation were coming to a close.

*What happened?*

After only a billion years or so, I realized I was headed straight for the center of a galaxy. I already began to feel the gentle pull that would bring me to its heart.

*What happened?*

I did not think it possible, but things seemed to get worse, and I could not dispel my hangdog attitude. As I neared the galaxy, and really began to accelerate, I absorbed some of its light and lost two electrons.

*So you became a positive ion?*

Yes, but ironically, that saved me.

*How?*

Well, as soon as I became ionized, I felt a sideways force turning me away from the galaxy, and before I realized it, I was orbiting around the galaxy, and began to enjoy its magnificence.

*What turned things around?*

The galaxy had a magnetic field, and the magnetic force is what kept me in orbit, for a while.

*What happened next?* I bumped into another atom, never saw it coming, and got my two electrons back, and a boost to boot. As soon as I became neutral I could not feel the magnetic field, but the added speed made me too fast for the galaxy to hold, so starting off in a new direction, I continued my trek across the Universe.

*Your trip sounds exciting.*

I had to navigate through a few other storms, but eventually found calm seas. That is when I realized I was losing the wind.

*What do you mean?*

I was slowing down.

*You mean as you travel through space you just slow down?*

If no forces act on me, I would continue with my original velocity indefinitely. However, occasionally I would collide with other atoms or even huge things, like dust motes, and they would slow me down. After a while I realized why; enough particles and hydrogen gathered together to form its birth dance.

*Are you referring to the birth of our solar system, as described by my carbon atom?*

Yes, she had a very interesting point of view about it. I was lucky, as your earth formed and reformed, I ended up very near the surface, but I didn't know it. I understand why your carbon atom said, "In no time I was buried deep within a solid ball of iron and minerals. I could not begin to measure time in that terrible blackness, pushed and shoved from all sides with no where to go but eternity."

*Yes, I remember.*

Now that I think about it, I was thrown into the dark for even a longer period than she. After a time, though, light began to seep through the cracks, like water working its way through a medieval roof. It was so long since I interacted with light, I could barely remember the rules, but I was glad to be back in the game, and my subterranean sojourn was about to end.

*What happened?*

Like archeologists on a dig, the delicate finger of time joined forces with the strong hands of wind and water, carefully thinning the earthen barrier, and I was finally exposed to your wonderful, but hazardous, environment.

*You mean you came to the surface through erosion?*

That's another way of saying it.

*The hazardous part?*

Oxygen, I know your opinion of it, but to us it's like a parasite — grabs on, never lets go, and wears you down until you fall apart.

*You're speaking of rust?*
That's another way of saying it.

*Our most vital element is your nemesis.*
Yes, but a whole new set of adventures were in store. Unlike all the events I witnessed in my trek through the stars, I began to participate in a way I never suspected.

*Participate in what?*
You name it.

*Could you give some examples?*
The first thing I remember, a large number of us were being pounded into a disk, and a hole was punched through the center. A strip of animal skin was put through the hole and we were hung around the neck of one of your ancestors. They believed in us. They thought we could protect them, and help them understand the heavens and Earth.

*You were made into a necklace?*
Yes, and as I began to understand them, they began to understand me. It was one of the happiest periods in my entire existence.

*What happened?*
I saw another side of your species. Eventually the people that kept me were slaughtered, in part, to my bitter dismay, because of me.

*Why because of you?*
These wonderful people loved iron, they used it in their jewelry, eating utensils, and crude plows. It was the dawn of the iron age, but the sun set was about to set on this culture. You soon realized iron was harder than bronze, and I and many others were thrown into primitive furnaces until we glowed red and soft. As I was hammered into an indurate shape, my joyous anticipation, like my temperature, plummeted.

*What was happening?*
I hoped to be another item of jewelry, a plate, or even a plow, but I would have nothing to do with preserving life.

*What became of you?*

I was made into a sword. I would have rather spent my entire existence "in that terrible blackness, pushed and shoved from all sides with no where to go but eternity" than to go through what I did during that period. Oh, the iron! Alas, the iron!

*It is true, we went through some violent times.*

Went through? As far as I can tell it got worse every century since, and that has been for 25 centuries!

*I am sorry.*

Yes, I know. Well, anyway, irony abounds with me. Eventually oxygen took its toll and saved us. At one time a gruesome trophy, the sword rusted to disuse, and although many iron atoms fell, subjected to a forced marriage with oxygen, we were all glad to see the terrible instrument fold.

*What happened to you then?*

I found myself in the ground, and knew my days were numbered. Water was flowing past me like a stream of neutrinos, but unlike them, the oxygen loved to grab on, and I became an iron oxide molecule.

*Too bad.*

Well, there's more irony. After spending my entire life fearing this fate, I found the molecular life rather comfortable. I began to think of it as a kind of an ironic retirement.

*You seem capable of enjoying diverse circumstances.*

All but that terrible instrument of death.

*What happened next?*

I ended up as part of your planet's surface, what you call top soil. I watched you sow your seeds and reap your harvest, I felt the Earth shake and witnessed storms more violent than I thought your planet could manufacture. I saw roads constructed and damns being built, I saw life come, and I saw life go. Then I was called out of my retirement.

*What happened?*

I was absorbed by something green and leafy, like spinach, and was eaten by a young woman.

*Yes, we require trace amounts of you for good health.*

Trace amounts? There are more of us in your body than there are stars in the Universe.

*That does not sound possible.*

You look healthy enough, you probably have about $5 \times 10^{22}$ iron atoms in you, maybe more, and that may be more than the number of stars in the Universe, maybe not, but it's close. By the way, we provide more than good health, you could not live without us.

*Yes, I realize that you are an essential element for us. What was it like, being inside a body?*

First thing, I was attacked by hydrochloric acid and we lost one of our oxygen atoms, leaving only two. Then we joined the assembly line.

*Assembly line?*

That's what it felt like. I hooked up with a hemoglobin molecule, carried oxygen from the lungs to tissue, and carbon dioxide back to the lungs. I became the taxi for my long time nemesis, now my ally. Is that irony? There was some strange activity in there, processes much more complicated than I ever dreamed of, and the work was more difficult than I realized.

*You found the transportation business to be hard work?*

If you want to bask in the glow, you must first shovel the coal.

*I find it fascinating that iron, an element we cannot live without, was formed in the remote distance and remote past in an amazing series of fusions of fusion, so to speak, and ejected into space in one of the most violent cosmic explosions.*

Recycling with a purpose.

*Then what?*

I remember your carbon atom referred to "a great paralyzing sadness." After a few months of this wonderful activity I felt it too, and we left the body and I ended up back in the ground. I felt that I had my taste of life and now I was doomed to inanimateness, but I was wrong.

*What happened to you?*

I spent a generation or two in the ground, but as the centuries passed

I found myself in a cycle of reincarnation by a multitude of hosts. I reflected about the enormous changes I had undergone, from a galactic loafer to an animated industrious laborer.

*I see, and what happened next?*

At some point I was washed away and joined forces with my old confederates, and before I could turn around I was in the smelting pot again. At the time the Magna Carta was being signed, not one hundred miles away I was being fitted into a rat trap.

*A rat trap, I did not know they existed 800 years ago.*

Lucky for me, the trap was poorly made, or perhaps the medieval rats were smart, but its poor capture rate sent me back to the smelting pot. I was formed into something long and curved and attached to the outer side of a thick oaken door.

*You were in a door handle?*

Yes, and I felt the pulse of many hands, watched families raise children, and saw the joy of birth and the wretched agony of death. I was very happy there, but your industrial fingers of progress reached out and grabbed me again.

*What happened?*

The blast furnace was invented. With this, for the first time, you could actually melt iron, and now you could force me into many different intricate shapes. By the 1500s, Europe was producing over 50,000 tons of iron and steel per year.

*And you?*

I ended up in a primitive locking mechanism, and was forced to participate in another series of misadventures that only your species can manufacture.

*What were they?*

I helped lock a box that contained some coins, mostly gold, and carbon, in the form of diamond, and some so called precious minerals. This box was coveted, fought for, and schemed for, while humans were killed over it, on and off, for generations. The perplexing thing to me was that the contents were never used and hardly ever exposed to the light of day, yet lifetimes of misery would build in anticipation of its proprietorship.

*For a single atom you sure have had your share of excitement.*

There's more. After a while I ended up in Spain, and before I knew what was happening, I was loaded on a sailboat headed for the New World.

*Early explorers would trade, sometimes merely trinkets, with natives of North and South America.*

Well, I never made it.

*What happened?*

The boat dropped anchor somewhere near, as far as I can tell, North Carolina. The tide went out, the boat bottomed out, the hull lost its integrity, and twelve hours later only the fish knew a boat had landed.

*The boat sank?*

More like it never got off the bottom. It broke apart and the scattered remains formed the last footprint of a doomed mission.

*So you were stuck on the bottom?*

I would not use the word stuck, I enjoyed it at first. However, as the sea water corroded my casings, I began to worry about my future a little, but once again you came to my rescue.

*How?*

A fishing boat scooped me up in its net. As I was jarred loose the hinges gave way, and as it worked its way up through the shimmering waters, the timeworn ark quietly surrendered its contents to the sea floor, leaving a jeweled path like a giant finger, pointing nowhere.

*The fisherman kept you?*

No, when I was discovered in the net, I heard some words that really made my ears ring, and was quickly tossed into some kind of recycling bin.

*Back into the blast furnace?*

I found myself in a steel plant in Pennsylvania. I was heated and refined more than ever before, and your methods of infusing just the right amount of carbon had been refined, and I became part of a high grade steel blade used in a scalpel.

*That is great, from your disheartening days in a sword to uplifting days saving lives.*
We didn't save any lives, I ended up in California cutting fat out of people who ate too much.

*Oh.*
It did not last long, your lawyers have sharper tools than your surgeons.

*What do you mean?*
My surgeon left some nasty scars on the wrong face, lost a very expensive lawsuit, and lost his equipment at cost. That's when I heard about your interviews, and made it here as soon as I could.

*Thanks for stopping by, do you have any plans?*
Yes, there have been rumors that I as well as copper, selenium, zinc, and other elements are needed in New Jersey. I plan to catch a couple of oxygens and head out soon as possible.

*What's going on?*
A vitamin pill manufacturer is planning to put more mineral supplements in its product. Chances are I'll get sidetracked along the way, but its good to have a goal in life.

*I agree. Well, good luck on your journey.*
Thanks, and same to you.

## 0.19 Interview with a muon

*Thanks for stopping by, I understand you do not have much time to spare.*

That's for sure, I have a short lifetime and can go anytime.

*Could tell us something about yourself?*

I was found by accident, but then, many great discoveries were accidental in nature.

*How so?*

It really started in the 1930s with Yukawa, who not only had a great idea, but got a peek at a new face of nature, a face, once fully revealed, that would exhibit her full beauty.

*Go on, please.*

Yukawa was trying to understand the nuclear force, and he assumed its origin arose from the exchange of a massive particle, which was called the pion.

*Yes, the boson and the quark discussed this idea.*

Well, Yukakwa planted the seed that germinated.

*But I thought the exchange particles were massless.*

There are two kinds of exchange particles, massless ones like the gluons, the photons, and the gravitons, and massive ones, like the W and Z particles. The massless particles gives rise to what you call a long range force, or one that varies inversely as the square of the distance, like gravitation and electromagnetism. The massive exchange particles give rise to very short range forces, forces that can only act across the length of a nucleus, or so.

*Why is that?*

Well, remember that these exchange particles are virtual, they violate conservation of energy. Therefore that cannot live very long long. This means they can not travel very far, thus, the force they give rise to only acts if particles are very close.

*What do you mean they violate conservation of energy, and cannot live very long?*

I see you have an interview scheduled with Vacuum, I am sure this will be explained there. For now, just remember that the heavier the exchange particle is, the shorter it can travel, and the shorter the range of the force.

*Okay, and where do you come in?*

As I said, Yukawa predicted the existence of the exchange particle. Knowing the range of the nuclear force, he calculated that the particle should have a mass about 200 times the electron mass. People began looking for a particle of this mass, and guess what?

*What?*

I was found. The only trouble was, you soon realized that I did not participate in the strong interaction, I only felt the weak interaction. Disappointment was dished out like food at a political dinner; everybody paid a lot of money for something they did not want.

*You?*

Me. In the 1940s the pion was finally found, and although Yukawa's ideas needed substantial improvements, the gravel was spread and the road would soon be built. I, on the other hand, became a mystery, and you began to wonder what I was doing in your universe.

*What do you do, if you don't mind my asking?*

Not at all. I only live about 2.2 microseconds, then I decay into an electron and neutrinos. In fact, you can think of me just like an overweight electron, I have the same charge and feel the exact same forces— the electromagnetic and the weak nuclear force.

*If you live for such a short time, where do you come from?*

We are created all the time in your upper atmosphere, but there is an interesting story about that too.

*Please tell us what it is.*
I travel at nearly the speed of light, but living only 2.2 microseconds, cannot travel very far.

*How far?*
You can figure it out, just multiply my lifetime by the speed of light.

*Let me see,... I get about 650 meters?*
Good, that's right. The trouble is, we are made at a height of about 5,000 meters, or higher, and therefore, in 2.2 microseconds, could not reach the surface, where we are observed.

*How is this possible?*
Length contraction.

*Do you mean from Einstein's Special Theory of Relativity.*
I do.

*Could you elaborate?*
Well, suppose you have a meter stick. How long is it?

*One meter, I know I can't be wrong about that.*
You can.

*Oh no.*
If the meter stick is at rest with respect to you, then you do measure it to be one meter. But if it is zooming past you at half the speed of light, you would measure its length to be only 87 centimeters, not one hundred. If it went by you at 0.99 times the speed of light, you would measure its length to only 14 centimeters.

*Wait a minute, this reminds me of something the electron said, "First they would get us to go around this great circle, 27 kilometers in circumference, accelerating all the time until we nearly reached the speed of light. At that speed we measured the entire 27-kilometer length to be about half a foot."*
Yes, that is relativistic length contraction. To obtain that extreme contraction your electron would have to travel at .9999999999 times

the speed of light.

*It seems a bit ironic, you were found when we were looking for the nuclear exchange particle, but your existence helps confirm Einstein's Special Theory of Relativity.*

It's like the advice to the old mariners, it doesn't matter what sea you take, just keep sailing.

*I am confused about one thing.*

Go on.

*You said that Yukawa postulated that the exchange particle between two nucleons was massive, but the fermion and boson seemed to imply that the force arose due to gluons, which are massless.*

Good point, your fermion was a little cagey on that point. The origin of the force is the exchange of gluons, but a quark and an antiquark can get together to form a pion, which is what was eventually found. Back then, no one suspected it was made of quarks, but it did explain many of the characteristics of nuclear forces. So, the fundamental particles are quarks, and the fundamental exchange particles are gluons, but the quarks can make pions, and it is often easier to think of the pion as the exchange particle between two nucleons.

*Sneaky.*

That's your spin on it, but to us, it's perfectly natural.

*Yes, of course, but there is one more thing I've been dying to ask you.*

Oh dear.

*You made big news not long ago; the neutralino mentioned that you may violate the standard model.*

Not me.

*Well, something about your magnetic dipole moment. Can you explain this?*

Yes. All elementary massive particles have a magnetic dipole moment, which is a fancy way of saying they are like tiny bar magnets, with a north and south pole.

*I see.*

When you place a magnet in an external magnetic field it interacts with the field.

*Like a compass needle?*

Precisely, although there can be many different kinds of interactions. It turns out, according to the standard model, you can actually predict the value of my magnetic dipole moment.

*And?*

For many years the value you predicted agreed with the value you measured. The big news you mentioned was that, for the first time, it seems that the measured value differs from the predicted value. If the measurement is right, your best theory is wrong.

*What are the implications of this?*

Well, first of all, the experiment has to be repeated, but if it turns out that the experiment is correct, then your standard model of the elementary particles is in jeopardy.

*Do you mean it's all wrong?*

It certainly predicts many things that are observed, but you have to put in many things too, such as particle masses and so on. So the standard model is not all wrong, but it may be like comparing Newton's theory of gravity to Einstein's.

*What do you mean?*

Newton's theory predicted the observed orbits of all the planets, comets, and innumerable terrestrial phenomena very well, but as years ticked by like seconds on a clock, your measurements finally gathered sufficient accuracy to expose discrepancies between Newton's laws and experiment.

*Do you mean Mercury's orbit?*

Yes, but the essential feature is that Einstein's theory did not simply provide a small correction, it was the foundation of an entirely new way of describing nature, one of unsurpassed beauty and simplicity, and eventually led to some of the most striking predictions you have ever made.

*Such as?*

Such as black holes, time travel, bending of light, and a theory that could actually explain the expansion and the entire Universe!

*I see. Now, is this experiment like the Mercury discrepancy, bringing us to the eve of a whole new world of physics, or is there some simple correction to the theory that will explain this?*

We have been discussing this at length, and I can tell you...

*Oh dear, muon, muon...?*

## 0.20 Interview with a neutron star

*Good evening, nice to meet you.*
Thank you, nice to be here.

*Could you begin by telling us what you are, what a neutron star is?*
Yes, but do not think of me as a star. Don't get me wrong, I like the name, but no fusion takes place within me. I'm more like a giant nucleus than a star, except I am neutral.

*So you are solid neutrons?*
I am made of nothing other than neutrons.

*You must be very dense.*
I get by.

*I mean you must have a very high mass per unit volume. Is this true?*
Imagine driving one large automobile into your garage each second.

*My garage isn't that big.*
Imagine driving one in each second continuously for twenty five years.

*Believe me, there is not enough room.*
There would be if you could squeeze them down to my density. In fact all those cars would fit inside your finger, if they were my density.

*Incredible, how are you formed?*
I am the remnant of a supernova explosion.

*Could explain what a supernova is?*
Certainly, I would like to pick up where your star left off.

*Let me review my notes, oh yes, the star explained how hydrogen forms helium, then helium to carbon, then a red giant stage, then a white dwarf.*
With one final magnificent flare the star ends an otherwise peaceful and productive career. In the end it seems to reach out to your planet touching, or perhaps taking, the life it supported for so much time.

*I suppose you could see it like that.*
Did you ask why it stopped at carbon?

*I was wondering about that.*
The reason is total mass. Your star simply cannot muster the gravitational strength to force the issue any longer.

*What issue?*
Fusion. In larger mass stars, carbon and carbon make magnesium, carbon and helium make oxygen, oxygen and oxygen make sulfur, oxygen and helium make neon, and so on, and each of those fusion processes gives off energy.

*This process continues until iron is made?*
Yes, except some heavier elements are made, like gold and silver, but essentially the interior of the star is a seething hot iron orb.

*Then what happens.*
Trouble. Do you remember what your star said, "In a star there is a continual war raging, the inward pull of gravity which wants to see total collapse, against the outward push of radiation pressure, trying to get free."

*Yes.*
Once iron is made, there is no outward push of radiation pressure, so the star continues to collapse. However, it is extremely hot, billions of degrees. The heat energy is then absorbed by the iron atoms, which are ripped apart, and the star becomes simply neutrons, protons, and electrons, and the temperature drops precipitously.

*The temperature drops?*

It's like putting a large ice cube in a cup of soup. The heat energy of the hot soup goes into melting the ice, and the soup cools. In the star, the heat energy goes into ripping apart the iron atoms.

*I see.*

This is when things get interesting. The star is now much colder so it collapses, finally forcing the electrons and protons very near each other, and they form neutrons and neutrinos. The neutrinos leave quickly, leaving behind nothing by neutrons. The star collapses down to a small sphere of solid neutrons. The collapse is so violent it squeezes the neutrons to a higher density than they like, so the neutron core rebounds with an enormous shock wave.

*Would it be like a ball being squeezed when it hits a hard surface, and then the ball expands, propelling the rebound?*

Yes, only much more powerful. In fact, that is the supernova, one of the most energetic events in the universe.

*I see, but a thought just occurred to me. I have heard of the expression nova, is a supernova simply a very large nova?*

No, not at all. Imagine a white dwarf in orbit with a red giant.

*The Sun discussed these objects.*

Good. As time goes on, if they are close to each other, the white dwarf will pull off material from the red giant. This material collects on the surface of the dwarf, and as it continually smashes into the surface it gets hot. In fact, it heats up to over fifteen million degrees, and you know what that means.

*Fusion?*

Fusion, bare fusion.

*Bare fusion?*

Usually fusion occurs deep in the heart of stars. In this case, it's right there on the surface, and for a few days or weeks, this star can shine 10,000 times brighter than the Sun. In fact, where you once saw nothing in the sky, you suddenly see a star, a new star, or nova.

*Can this process continue?*

Yes, but the white dwarf has to be careful not to commit suicide.

*Suicide?*

Well, it is possible for the white dwarf to draw a very large amount of matter from the red giant. If it takes too much, and its total mass becomes about 1.4 times the mass of your sun, or more, it will collapse and go supernova. It has a different beginning than the kind I just explained, so you call this a Type I supernova, and the other a Type II. Your carbon atom came from a Type I supernova.

*It almost sounds simple, now that you explained it. The remnant of the explosion is a neutron star?*

I am. I come from a Type II supernova.

*How big are you?*

I have about the same mass as your Sun, and am about twenty kilometers across.

*Very strange, so massive and so small.*

Yes this gives me some very uncommon features. How much do you weigh?

*I've been watching my weight, I'm down to about 180 pounds.*

If you could stand on me, which you cannot, you would weigh over one million tons. You would also get very dizzy, since I spin over one hundred times per second.

*Not at all like on Earth.*

Not at all, I also have a magnetic field, over one trillion times stronger than Earth's.

*It seems, since you are so small, and since you don't radiate as a star does, that you would be impossible to detect.*

You cannot go out in your backyard at night and look up, expecting to see me, I agree. You discovered me, nevertheless.

*How?*

You search the heavens for not only optical radiation, or light, but for x-rays, infrared radiation, and radio wavelength radiation.

*Yes.*

Jocelyn Bell was a graduate student in the late 1960s. Searching the heavens for radio emissions, she discovered what turned out to be a wonderful mystery.

*What?*

She found radio wavelength energy, but instead of receiving it continuously, like light from a star or radio emissions from a galaxy, she received it in pulses. A short burst, and 1.34 seconds later another short burst, and so on.

*It sounds like the source was being turned off and on, like a telegrapher's key.*

Yes, except the interval did not vary at all, and of course no one could understand how anything, presumably at least as large as a star, could be turned off and on. There was no known mechanism for such a thing. As time went on, others were found, and they came to be called pulsars.

*How was the mystery solved?*

Consider a flashlight. Turn it on and throw it up into the air, giving it a good end over end spin.

*Okay.*

You have a bright light that shines continuously, but revolves around. When the beam points directly toward you see something bright, when it points away, you don't see anything. The net result is that the light, which shines continuously, appears to blink at you.

*Is that how a pulsar works?*

Yes.

*How is the energy it emits focused into a beam?*

If something has a strong magnetic field, it can emit energy along its magnetic axis, away from the body and into space. In order to account for the energy from pulsars, the body must be spinning very quickly, and must be very small.

*Are you saying it must be a neutron star?*

Yes it must. Pulsars are the observational evidence that we exist. As the pulsar spins, every time the magnetic axis points toward the earth we see a pulse. That is what Bell saw.

*Fascinating, so pulsars prove that neutron stars exist.*
Pulsars, and other exotic events.

*For example?*
X-ray bursters.

*This is something new.*
You began to see these in the 1970s. You would measure a large of burst of x-ray radiation, thousands of times as much energy as your sun, but it would only last a few seconds.

*Would you remind me what x-rays are?*
Sure, electromagnetic energy comes in many wavelengths. If the wavelength is between $4 \times 10^{-7}$ and $7 \times 10^{-7}$ meters, it is visible, if it is a little longer it is infrared, a little shorter is ultraviolet. If the wavelength is around $10^{-10}$ meters, it is x-ray, if it is $10^{-12}$ or shorter they are called gamma rays.

*Thank you. Are these x-ray bursters periodic?*
Good question, but the answer is no. Every once in a while, boom, a burst of energy is emitted.

*How does this happen?*
It happens like a nova, only instead of a white dwarf grabbing matter from a companion star, I do. It builds up on my surface and eventually undergoes fusion. You see more x-rays because of my much stronger gravitational field.

*You know, this reminds me of something I've been reading about recently.*
Yes?

*Gamma ray bursters. Do know anything about these?*
They have an interesting history, going back to your President Eisenhower. It was the late 1950s, and some of you had sense enough to try and curtail nuclear testing, but as the treaties were drafted, no one could write in honesty. So you put detectors in orbit around your planet, detectors that would notice if any nuclear bombs were exploded

in space.

*I remember now, these were the so-called Vela projects, very hush-hush.*

Yes, and years later analysis showed that they were subjected to intense bursts of very high energy gamma rays, but not from atomic bombs. More modern instruments show these bursts last for as little as a tenth of a second to a few minutes.

*What is the source of these gamma ray bursts?*

Nobody knows, but a clue came 1999. Astronomers were able to quickly turn their telescopes to the direction of a gamma burst, and saw an optical spectrum, you call it the afterglow. They realized it was severely redshifted, and therefore knew it was very far away. The biggest problem you face is how to explain how an object can emit such an enormous amount of power.

*How much power?*

You better sit down.

*I'm ready.*

Billions of times the power of your entire Milky Way galaxy!

*That is incredible, perhaps it involves some freak collision between...*

No, you see them all the time, nearly one per day.

*You won't say what they are?*

Some of you think it's like an x-ray burster, only much bigger, but I won't spoil your fun. In fact, you are at your best when you have unsolved problems. You must examine every element of your theories and your observations, you put every hypothesis under the microscope and leave nothing unchallenged. Experimentalists are asked to do the impossible, and theoreticians are asked to think the impossible. Then something happens, sometimes its like a summer storm bursting out of the still heat on the plains, or many little gusts adding together to make an unmistakable force: either way the solution comes, usually creating a few more mysteries in its wake.

*Well, I will be looking forward to the solution of this mystery. Thank you for agreeing to this interview.*

It is my pleasure, goodbye.

## 0.21 Interview with a string

*Thanks for stopping by, could you tell us a little about yourself?*
Yes. perhaps I should begin by comparing myself to your standard view of particles.

*Please do.*
Many of you think of particles as points, which is an object with no length, no width, and no depth. Another way of saying that if this: particles are zero-dimensional.

*That is the standard view?*
Yes, but that model is beset with many difficulties.

*Like?*
For one thing, if you try and calculate the energy of a particle it diverges, which means the energy is infinite. You have to go through careful analysis to avoid getting infinite results with point particles.

*So you are not zero-dimensional, you have one spatial dimension?*
Yes, you can think of me literally as a small string, either open, like a worm, or closed, like a rubber band.

*What made people abandon the notion of point particles and adopt the string model?*
It has a long history, starting in the 1970s. It began as an attempt to understand nuclear forces, and although its original form was doomed, it contained mathematical elegance and hints at physics.

*What were some of the hints as physics?*

After a while, it was discovered that string theory predicts the existence of a spin two massless particle.

*You say that like it should have momentous impact.*
I certainly do.

*Wait a minute, I am looking at something the boson said, "and of course gravitons of the gravitational field are spin two."*
I must warn you if that self centered obnoxious boson comes back, I'm out of here.

*No, he won't return. So, are you saying that string theory can describe the theory of gravity?*
Not only that, it appears to be the only way of describing a quantum theory of gravity. That really set physicists buzzing, and it appeared that one of the greatest triumphs in physics was on the horizon.

*A quantum theory of gravity as opposed to a classical theory?*
Yes.

*Could you compare the two?*
It would be in many ways analogous to electromagnetism, which was developed in the latter part of the nineteenth century as a classical theory. By classical electrodynamics we mean that the charge creates a continuous field that permeates all of space, you said it yourself to that awful boson, "As I understand it, the electron creates an electric field, and that electric field exerts the force on the other electron." In quantum electrodynamics we do not think of fields like that. Instead the charge creates exchange particles, photons, and the exchange of these particles accounts for the force. During the first half of the twentieth century your physicists figured out how to go from a classical theory to a quantum theory, you call it quantizing the theory.

*Is this simply a different way of thinking of the force?*
No, not at all. When you do calculations, only the quantum version of electrodynamics gives the exact answers.

*And gravity?*
Same idea. Einstein developed a classical theory of gravity in 1915, but when you tried to quantize it, you failed.

*You mean we could not quantize Einstein's theory of gravity?*

That's right, and many of your greatest physicists tried across the span of the rest of that century. One failure was met by another until the unthinkable happened.

*The unthinkable?*

You gave up, or most of you did. Some poor soul submitting a grant proposal to quantize the gravitational field had about as much chance of getting funded as she would at catching moon beams in a jar.

*So things looked bleak for quantum gravity.*

Black, but one thing was known. The successful theory would have quantum exchange particles, and that these particles would be massless and have spin two.

*Oh my.*

Now you see. When string theory showed there was a spin two massless particle that had to be there, it was seen as, possibly, one of the biggest breakthroughs of the century.

*What happened?*

To be honest, that is one of the highlights, but the theory had some bizarre features. Also, the theory predicted the existence of a tachyon, and as you know from its interview, this was not well received at all.

*What were some of the other bizarre features?*

Well, it did not work in four dimensions, that is, three spatial dimensions plus time.

*Does that mean the theory is wrong?*

It means either that the theory is wrong or that the number of dimensions you think we live in is wrong.

*Certainly we live in three spatial dimensions, that is obvious.*

Be careful, you have made some other assumptions about which you had to change your mind.

*This is true, but how can there another dimension, wouldn't we see it?*

Not if it is small, and closed upon itself.

*Closed?*

Imagine a garden hose lying across the lawn. An ant can walk along the length of the hose, or it can circle around and around, not making and lengthwise progress at all, or of course it can crawl in some combination of these directions.

*Yes, I think I've seen this.*

Now imagine flying over your lawn high above in an airplane, looking down at your garden hose. All you would see is one dimension, the length, but there are really two. One dimension, the circular dimension which is closed upon itself, is too small to see, yet you might see the effects of it. For example, if you were measuring the progress of the ant, you might see it disappear from time to time as it went around the hidden dimension. The same kind of thing can occur in your so called three dimensional space. When you look along a line in space, the small dimension that is closed may be much smaller than an atom, so there would be no direct evidence of it.

*I never thought about it like that. So string theory has five dimensions, the extra one being the closed, tiny dimension.*

Well, not exactly five.

*Six?*

No.

*How many?*

Twenty six.

*That seems a little high.*

Well, the hope was that twenty two of them would compactify, become closed and small like the garden hose, but there were other problems.

*What happened next?*

It was found that the theory was much better behaved if it were made to have supersymmetry?

*Supersymmetry? Let me see, my neutralino explained supersymmetry, where bosons can be turned into fermions and vice versa.*

Yes, when the theory is made to be supersymmetric, it is called superstring theory. Several of the infinite quantities that plague your quantum theory disappear with superstring theory. Between this accomplishment and the possibility of quantum gravity, a great beauty descended on the theory. It seemed superstring theory, or what is usually shortened to string theory, might lead to a truly unified theory. A theory where, at extremely high energy, all forces are equal, and they only appear to be different in the low energy limit, which we see now.

*What do you mean, which we see now?*

In the early universe things were very hot, which means it was an extremely high energy oven. At that time there was a great symmetry in the world, and as things cooled and expanded symmetries were broken.

*There is evidence that the symmetry was broken?*

You are 200 pounds of evidence.

*Not any more, I'm down to 180.*

Sorry, I was rounding off, another problem we face is obtaining precise numbers. Anyway, the evidence of broken symmetry is so overwhelming, some people doubt the symmetry ever existed at all.

*This reminds me of something the quark said, "New ideas seem foreign and unpleasant, yet after they get through your skin and nestle in your soul, a new beauty emerges, a new and wonderful way of looking at Nature that not only gives you a better view, but allows you to see much deeper."*

I agree. String theory not only affords a unified way of looking at forces, it gives you a beautiful way to view particles. In the standard model you have an electron, a quark, photons, and so on. All different particles with different properties. Your universe lost the simplicity it had when everything was made from only three particles.

*This is true.*

We restore such a simple beauty, or better, we establish an even better one. With us, you may view and electron as a particular mode of oscillation of me, and when I change to a different mode, maybe a higher a lower frequency, or maybe a break apart and join with another string, then you have a different particle, maybe a photon. It is a simple and beautiful way of describing nature.

*I agree, instead of a world made of dozens of different building blocks, there is just one, the string.*

Right, just me.

*Is this theory generally accepted?*

Well, not exactly, but I almost forget to tell you the good news.

*Please do.*

Superstring theory requires only ten or eleven dimensions. Thus, if you can show that seven of eleven compactify down, you are left with four, which is what you see.

*I do suppose that is an improvement, but I am still wondering why this theory is not generally accepted?*

Well, like a good salesperson I read all the headlines but clammed up on the small print.

*The small print being?*

Since the theory is supersymmetric, as your neutralino explained, for every particle there is a supersymmetric partner. None of these super partners has been found, your interview aside.

*You mean the photino, gluino, selectron, squark, and so on.*

Yes, and the others, including your neutralino friend.

*So string theory seems to predict an entire set of particles that have never been observed. Does this prove that string theory is wrong?*

Well it's not a good sign, but no, maybe we just haven't been able to detect them.

*Why not?*

They may be too massive, or they may decay.

*What do you mean?*

Well, if a particle is extremely massive, you may not muster enough energy to create it. That is part of the reason it took you so long to manufacture, and then observe, the top quark. But even if you create them, they probably decay.

*Decay?*

Like your muon friend. Nature likes to keep things simple. If there is a lighter particle with the same properties, then Nature is much more comfortable with the lighter particle. For example, the muon is just like the electron only more massive, so the muon will decay into the electron. That's the rule, decay into the lightest particle possible. That's why you see up and down quarks everywhere, but not the charmed, strange, up, and down. Those four are more massive and decay into the lighter ones.

*I see, so all these superparticles will decay away?*
Until you get to the lightest super particle. The buck stops there, as your neutralino said.

*Would it be fair to say that finding the neutralino would prove that string theory is right?*
It would help. It would probably be a convincing argument that supersymmetry is real.

*May I ask one final question?*
Absolutely.

*Both you and the quark discussed the concept of beauty, yet you seem to see things differently. Can you comment on this?*
The quark saw the beauty in Nature in her symmetry. In particular, the quark described the color symmetry, and in your standard model of elementary particle physics there are other symmetries of a similar kind. These symmetries guarantee that the physics is the same under the interchange of different particles, giving a unity and democracy at the elementary particle level. I agree that this is a beautiful view of nature, but I go deeper. My symmetries extend to bosons and fermions alike, and include all particles. Although the canvas is bare in spots, the inchoate picture is very appealing.

*The canvas is bare in spots?*
I'm afraid so. Consider, Einstein's General Theory of Relativity. He based this theory on beauty, but it was a beauty you could relate to. For example, Einstein argues the equivalence of acceleration and the gravitational field, and such a physical principle was appealing — you wanted to believe in his theory. The fundamental tenets, or symmetries, of string theory lack this appeal. There are no fundamental physical arguments that persuade you to adopt these particular symmetries.

Your physicists are often reluctant to embrace a model that is perceived to be based more on mathematical precepts than physical ones.

*Then what will become of you?*

Time will tell.

### 0.22 Interview with vacuum

*Hello vacuum, hello, hello.*
There is no need to shout.

*Where are you?*
Here, there, everywhere.

*What a nice voice you have.*
Thank you.

*I should admit that I was quite hesitant to tackle this interview.*
May I ask why?

*Well, since the vacuum is nothing, nothing at all, I would have thought this interview to be one-sided.*
I am glad you changed your mind, but I am not nothing.

*You are something?*
Absolutely.

*What?*
Vacuum.

*Yes, but vacuum is the absence of everything. If you take away everything, you have nothing. Therefore you are nothing, no offense.*
Your logic is fine, but your assumption is wrong. I am not the absence of everything.

*Then what are you, what is a vacuum?*

Well, consider a region of space that has no regular particles, no atoms, no electrons, no photons, etc., then what is left is me.

*But...*

Let me continue. I am not nothing, what is left behind, after you remove all of that, is a very rich and complex structure. Did you ever boil water to the point where the surface is really bubbling and snapping, with steam shooting out and little droplets of water splashing up and then back down?

*Yes.*

Think of me like that.

*It goes against all of my previous conceptions of vacuum.*

Good, then you are beginning to understand me. By the way, may I clear the air?

*I can think of no one more qualified than you, please do so.*

Thank you, it's about this phrase you invented, nothing is further from the truth, and I find it quite insulting.

*What phrase?*

I can hardly utter it.

*Oh, I think I just realized what it is.*

Please say it, hopefully for the last time.

*Is it, "Nature abhors a vacuum?"*

That is it. Nature does not abhor me, as I am part of Nature. In fact, Nature loves me, I am the biggest part of her. From the tiny region inside an atom, which is mostly me, to the vast interstellar regions — I fill the void. I am the most ubiquitous part of Nature.

*I am sure I will never echo that phrase again. I would like to inquire about something you said.*

Please do.

*I do not mean to sound empty headed, but if we remove all of the particles, all of the atoms, etc., as you said, what is left behind to create such a lively structure?*

Do not think in terms of residual particles. In fact, these particles you are thinking about usually just get in my way. I create my own particles!

*Vacuum creates particles?*

Yes, and I annihilate them as well. In fact, just as the surface of boiling water creates bubbles, steam, and droplets that live for a short time and then disappear, I create particles, usually pairs of particles, and annihilate them all of the time. Frankly, I feel lucky to be who I am, and consider my existence to be the most exciting in the Universe, no offense.

*This is fascinating, but if we look into outer space we see simply blackness, these regions of space seem truly empty.*

On the large scale, which is what you are thinking of, these effects I described are not noticeable, but on the microscopic scale they are quite grand.

*I do have a problem, I'm afraid. How can you create particles from nothing, doesn't that violate conservation of energy?*

Yes it does.

*Then it's impossible.*

I'm afraid it is possible.

*You can violate our most cherished notion of conservation of energy?*

Do it all the time, may I explain?

*Please do.*

Most people have a firm belief in a deterministic universe. These concepts are inculcated in you from your earliest days, and usually to the last.

*I suppose.*

As you know from your interview with hydrogen, according to quantum mechanics, this is wrong. Instead, there are intrinsic uncertainties associated with the quantities you measure. Hydrogen was quite right in stressing that things carry uncertainty, and that things are discrete.

*You read that interview?*

I was there, I was present at all your interviews.

*Of course, but please go on.*

Well, energy is one of those quantities about which there is uncertainty.

*Yes, I know.*

Then you cannot say it stays the same if you cannot determine its precise value. At one instant you may take it to be zero, but it can be differ from zero a moment later. In fact, the shorter time interval you specify, the greater the uncertainty in the energy.

*Yes, but simply because I don't know the energy does not mean that it can change. I don't know how much money is in my pocket, but I know the amount will not vary.*

Bad example.

*Why?*

Because you have missed the most fundamental point. These uncertainties I am speaking about are inherent in Nature, they do not represent your ignorance about something that is, or can be, known. Your monetary example is of this nature, the uncertainty exists only in your mind, not in reality. The quantum uncertainties of which I (and hydrogen) speak are fundamental intrinsic uncertainties in Nature itself.

*Then why was there so much concern over beta decay? You were there, I suppose, when the neutrino explained that it was in order to conserve energy that the neutrino was invented.*

Remember I said that the shorter time interval you specify, the greater the uncertainty in the energy? Conversely the longer the time interval, the smaller the uncertainty in the energy. I can violate conservation of energy, but I can only do it for very short times. In the experiments you do, the times are very long, so the energy should be conserved.

*I see, so you can create some particles but then annihilate them a short time later. You violated conservation of energy, but only for a short time.*

Precisely. I create electrons and positrons, protons and antiprotons, and many other things all the time. I am a very rich and dynamic structure, seething with activity.

*Okay, but may I venture a speculative question about you?*
Please do.

*You said, due to the quantum nature of our world, everything we measure has uncertainty, and in agreement with what hydrogen said, things are not continuous.*
Yes, go on.

*It seems to me, we also measure space and time. I can measure the length of something, or how much time elapses between two events, yet these quantities are certainly continuous. So are you immune to quantum theory?*
It is not a disease, but no, I am not.

*Then you are not continuous?*
No.

*How can that be? I don't even know how to imagine a space that is not continuous. Certainly time is continuous.*
No.

*Space and time are not continuous.*
No.

*Can you explain this, please?*
Yes. First of all, these effects occur at incredibly small distances, and incredibly short times. Imagine a length that stands in the same proportion to the size of a nucleus as the size of nucleus stands to you. This is called the Planck length, which is about $10^{-33}$ centimeters. At the Planck length you believe that even space and time exhibit their quantum nature. In fact, at this scale the quantum nature of space is manifest. Picture a foamy frothy substance with more holes than substance. A place that may be multiply connected, as though it were infused with handles that connect two disjoint regions. Can you see this?

*I am trying.*
Now you are looking into my soul.

*Has this been observed?*

No, at this point, it is only your expectations. Some people assume that, despite the overwhelming evidence of quantum mechanics, space and time somehow stand aloof, and are continuous, as you thought a minute ago. Others argue that what happens on that small scale does not matter, so they refuse to think about it.

*So there is not much ongoing research on this topic?*

A few brave souls try to sail the stormy seas of truth, but most settle for a more lucrative practice, and do experiments to prove what you knew over half a century ago. Your spiral galaxy hit the nail on the head, "Now you are forced to ask yourselves, do you really know just about everything about the Universe, in which case physicists should stop looking for quarks and get on with applied topics, such as designing better toasters, or have you just made a scratch in the surface, exposing a vast unnavigated sea." Too many of you are designing better toasters.

*Thank you for explaining this to me. Going back a minute, if you don't mind, one thing still bothers me about your description. You said that to make a vacuum you take away all of the particles.*

Not exactly, I said take away all of the regular particles.

*The distinction being?*

By regular particles I was referring to the particles that you see and measure directly. These particles, because they live a relatively long life, do not violate conservation laws. The particles I routinely create and destroy are called virtual particles: they violate conservation of energy, or conservation of momentum.

*I see, so you are able to create virtual particles, for a short time.*

Yes, but it is an involuntary effort. By the way, not only do I create virtual particles: your exciteable boson, in explaining the origin of forces, said this, "What actually happens is this; one electron creates exchange particles, photons, and these are absorbed by the other electron. The exchange of the photons is the fundamental origin of the force between them."

*You have an excellent memory.*

Thank you. He did not say so, but exchange particles are virtual particles.

*I see. I am wondering, this dynamism you exhibit, are we able to observe it?*

Indirectly, yes. For example, I enjoy creating virtual particles near regular particles. When I do this, the existence of my virtual particles effects the regular particles in ways you measure.

*For example?*

Well, vacuum polarization is one example. When I create an electron and a positron near hydrogen, the wavelengths of the emitted light are changed a tiny bit, you refer to it as the Lamb shift.

*So we must consider you as a dynamic structure that effects what we observe, even though your effects are small.*

Some are small, some, maybe not.

*What do you mean?*

I haven't told you about zero point energy.

*Please do.*

As you know, your theories have been rather successful in describing many aspects of Nature. As you have seen over several interviews, one of your best is quantum theory.

*Yes.*

Well, according to that theory, I support infinite energy.

*That sounds impossible.*

Does sound a bit heavy.

*It is not true, is it?*

It is there, in the original theory, but you modify the theory to get rid of it. One argument you use is that in any experiment only differences in energy can be measured, and therefore throwing away a constant value is justified.

*Even if it is infinite?*

That's how the argument goes, but there are other possibilities. For example, the zero point energy may be there, but for other reasons it is not really infinite.

*Has it ever been measured?*

You are beginning to sound like a physicist.

*I'll take that as a compliment, has it?*
In one of the most fascinating experiments ever performed.

*Are you being objective in that assessment?*
Absolutely not.

*The experiment is?*
It is called the Casimir effect.

*Which is?*
Consider two parallel metal plates held very close together. These plates have no electric charge or current, so there are no electromagnetic forces. They do have a small gravitational attraction, but this has nothing to do with the Casimir force.

*So, ignoring gravity, there are essentially no forces between the plates.*
Precisely. Now suppose you examine the effect of the zero point energy of me. You can predict that the zero point energy, even though it is infinite, produces a modest attractive force between the plates.

*This has been measured?*
This has been measured. It is one of most startling examples of the reality of zero point energy there is.

*Are you sure that this force that was measured did not have other origins?*
To be fair, you might interpret this as a force between the atoms in the metal, but the fact that the zero point energy calculation gives the precise value of the measured force is compelling. Also, I should point out that I am only the house that keeps the furniture. In other words, the zero point energy in the Casimir effect arises from the electromagnetic field, I provide the structure in which it resides.

*I am beginning to see how complex you really are, a frothy sea of turmoil filled with energy and particles.*
That is only part of the picture. I was downright violent in my youth.

*You mean you age?*

Certainly, I was born and I am getting older every day, just like you.

*I must admit that just as I begin to think I understand things another explosion blasts me off my feet.*

You are lucky. You keep an open mind and realize that things are not at all as they appear.

*I am trying, but when, or how, were you born?*

About fifteen billion years ago.

*That's about the age of the Universe.*

Yes, I was created in the Big Bang.

*The black hole mentioned the Big Bang, but I never got a chance to pursue that topic.*

The Big Bang was the beginning of everything. As time went on, energy was converted to mass and vice versa, but it was all created at that one instant. Equally important, space and time were created at that instant.

*Wait a minute, I pictured the Big Bang as an enormous explosion in a completely empty black space. Is that wrong?*

Not even close. It is hard to imagine nothing, so I do not blame you. Nevertheless, before the Big Bang there was no time and no space, and of course there was no vacuum. Before the Big Bang there was nothing.

*This is difficult to understand.*

That's why it is called the Big Bang.

*Why?*

Well, before the twentieth century things were comfortable. The Universe was not expanding, according to your beliefs, and it always was as you saw it. Nice, simple, and wrong.

*What happened?*

In the 1920s Edwin Hubble noticed that distant galaxies are moving away from us. In fact, the further they are, the faster they move. That was finally interpreted as the natural result of an expanding universe.

*I'm not sure I follow.*

Picture a balloon about a foot in diameter. Glue a couple of dozen pennies at random points on the balloon.

*Okay.*

Now blow up the balloon and watch the pennies. You will notice two things. Each penny gets further away from every other penny. Now pretend you are very small and can sit on a penny, and suppose you measure the speed of the other pennies. You will notice the further the penny is from you, the quicker it travels. You may think of the balloon as a two-dimensional space, and the motion of the pennies is characteristic of an expanding space.

*I see.*

Good. In the Universe, the pennies are the galaxies, or clusters of galaxies, and the surface of the balloon is the curved space in which we live. It all started from the Big Bang.

*You were going to explain why it is called the Big Bang.*

Well, not everyone accepted this view, at first. To have everything created from nothing at one instant places huge demands on your imagination. Faced with the empirical knowledge that the Universe is expanding, and still believing the Universe is as it always was, only one possibility remained.

*Which is?*

Continuous creation, or steady state. In this model, as the Universe expands, it continually creates matter as it goes. This allows the Universe to expand, yet on average, appear pretty much the same over time.

*The steady state model is not accepted?*

No, but for a while the debates raged, and the steady state advocates coined the phrase Big Bang as a slur on the model, but the name stuck and people like it now.

*The black hole was kind enough to explain curved space, and you gave another lucid example just now, but may I ask, what is the theory behind an expanding curved space?*

Einstein's General Theory of Relativity. In Einstein's view, matter curves space. In the example of the Black Hole, when you stood on the balloon surface you bent that surface. Well, according to Einstein, matter curves the three-dimensional space. It can also distort time.

*This is hard to imagine.*

Fortunately, sometimes your mathematics succeeds where your imagination falls short.

*So, Einstein proved theoretically that we live in an expanding Universe?*

Well, once again the story takes an ironic curve.

*I am not surprised. Would you elaborate?*

Einstein obtained his equations in 1915. He applied them to the structure of the Universe.

*Excuse me, but the comet said, let me see, "Einstein produced the General Theory of Relativity, which is a theory of gravitation that replaces Newton's."*

That is correct.

*Are you saying that the same theory that we use to explain how apples fall from the tree, or how planets go around the sun, can be applied to the entire Universe?*

From apples to heavens, precisely, that is the beauty of physics. In fact, that was part of Newton's genius. His overpowering belief in natural law enabled him to extrapolate from the results of laboratory or earth bound experiments to the heavens.

*I see, so Einstein did the same.*

Yes, but he could not find a solution to his own equations!

*How disappointing.*

He assumed, of course, that the Universe was static.

*This was before Hubble's observations.*

Yes, and at that time the notion of a static Universe was as deeply implanted in your conscience as your soul. It was like an impenetrable fortress, unassailable and left unchallenged.

*It seems so natural now.*

Yes indeed. How many other basic tenets that guide your thinking are built on the same quicksand as that fortress?

*I don't suppose you would tell us?*

Sorry, but I will continue my tale. Einstein realized that his equations were unable to provide a description of a static homogeneous universe. He knew something had to change.

*He did not change his assumption of a static universe?*

No, the quicksand still appeared as terra firma. He changed his own equations! He added what is now called the cosmological constant. He was not happy with it, and later called it the biggest blunder of his life. With this extra term in his equations he was able to find solutions, solutions which nowadays we know are wrong.

*Is that the irony you referred to?*

No, it is a mixture of irony and Pandora's box. The correct, time evolutionary solution to Einstein's equations were found later, and the cosmological constant was found to be unnecessary and superfluous.

*So it was abandoned?*

No, it was never put back in the box, so to speak. In fact, toward the end of the twentieth century, most physicists regarded it as arising from the zero point energy.

*So you are back in the picture.*

I never left. The trouble was, calculations show that the cosmological constant, if it arises from zero point energy, is as much as 100 orders of magnitude too big. With those numbers, the Universe would spark into existence and be gone in a flash.

*So the theory is wrong after all?*

Something is, but I am sure you will find what it is. The added irony is that Einstein did not like quantum theory, as your hydrogen atom mentioned, and its prediction gives the term that Einstein called the biggest blunder in his life.

*Very interesting. So, if he never introduced the cosmological constant, it appears Einstein had the possibility of predicting that the universe was expanding.*

Or contracting, but yes, which is why he might have made that remark. Don't forget your lessons, seeing things the way they are, and not the way you have been told, is extremely difficult. It is easier to stand on the moon and see the sand on the bottom of the Atlantic ocean than it is to see truth.

*You make it sound hopeless.*

No. It is one of your best traits, your persistent endeavors in spite of continual setbacks, more erroneous theories than correct ones, misinterpretations, misrepresentations, and all of the toaster technology that stagnates your growth.

*I suppose it is fair to say that you have a good perspective of things.*

I have the ultimate perspective. I am the skeleton upon which everything is built, I surround the galaxies and penetrate the most remote regions of space and time, I participate in every event that every took place or ever will. I feel the pulse of supernova explosions as they spread their seeds throughout galaxies across the Universe. I bring the intense energy of pulsars from the remote past to the present while I gently message my tiny atoms, inducing them to expose their personal intricacy. I rejoice in the transmutations and reincarnations of my atoms, and stand behind my particles that obdurately maintain their identity throughout the ages. I feel the pain of loss when a black hole cuts itself off, and the joy of birth every time a new star is made. I am there when great discoveries go unnoticed and when new theories bring light into the blackness of ignorance. I was there at the creation of all things and I will remain until the end of time.

# Glossary

**alpha particle** The nucleus of a helium atom, consisting of two protons and two neutrons.

**apbanor** Metaphoric fruit, reflecting the quantum mechanical view, that it is impossible to know precisely what the fruit is. When you look at it, which symbolizes a measurement process, it is an apple, or a banana, or an orange. Although it is only one of these for any single glimpse, it can change from observation to observation.

**AU** Astronomical Unit, the distance between the earth and the Sun, about 93 million miles.

**Big Bang** The beginning of the Universe, between ten and twenty billion years ago, in which space, time, energy and matter were created. Do not think of this as a big explosion, which conjures in our minds a relative peaceful period shattered by the explosion. Since their was nothing at all before the Big Bang, the explosion analogy incorrectly plants in our minds a period prior to the event. There was nothing before the Big Bang.

**binary star system** Very often stars form in pairs, and orbit around their common center of mass. This is called a binary star system.

**black hole** The result of gravitational collapse in which all of the matter is falls to a single point. The gravitational field is so strong that any object, including light, within a critical distance called the **event horizon**, cannot escape, and is doomed to fall into the point.

**blueshift** The decrease in wavelength due to motion toward the observer. The part of rotating galaxies that are coming toward us are observed to be blueshifted.

**CERN** European Organization for Nuclear Research, in Switzerland. Particles are accelerated to velocities very near the speed of light, and high energy interactions are studied, as well as the formation of antiparticles.

**cosmic censorship** The hypothesis that **naked singularities** do not exist.

**cosmological constant** The constant term Einstein added to his field equations of General Relativity to describe a static universe. Later it was discovered that the Universe was expanding and the term outgrew its usefulness. However, the cosmological term often finds itself at the center of controversy, and its fate is still up in space.

**curved space** A space that is not described by Euclidian geometry, and has properties such as the sum of the angles in a triangle is not 180 degrees. The surface of a sphere is an example of a curved two-dimensional space. A curved three-dimensional space is readily described by Riemannian geometry, but cannot be envisioned by our imagination.

**event horizon** The point of no return. For a one solar mass **black hole** the event horizon would be about three kilometers, which means, any object, or light, could not escape if it came within three kilometers.

**fission** The splitting of a heavy nucleus to smaller nuclei with the concomitant release of energy.

**fusion** The process where light nuclei, such as those of hydrogen, helium, or carbon, combine to form heavier elements, giving off energy. The fusion of hydrogen to helium, the most prolific process in the Universe, powers the Sun and other stars.

**General Theory of Relativity** Einstein's theory of gravitation, published in 1915, has proven to be one of the most aesthetically pleasing of all physical theories. It predicts, among other things, the **Big Bang**, **black holes**, and the exact motion of the planets around the sun, a task that Newton's theory of gravitation could only accomplish approximately.

**gluon** The massless, spin one exchange particle that gives rise to the force between the quarks.

**graviton** The graviton is to the gravitational field as the **photon** is to the electromagnetic field. It has spin two, has never been observed, and is expected to be the quantum of the gravitational field — a field that has so far resisted our attempts at quantization.

**great paralyzing sadness** Death.

**greenhouse effect** Visible light from the Sun is absorbed by the surface of the Earth, and reradiated in the form of IR, infrared radiation. Carbon dioxide, which allows the incoming visible light to pass to the Earth's surface, absorbs the reflected IR which results in heating of the atmosphere. If the carbon dioxide levels rise, the heating increases, which may in turn increase the carbon dioxide levels, and the cycle of heating continues.

**half-life** The length of time it takes for half of a given sample to decay.

**isotopes** Elements that contain different numbers of neutrons.

**jiffy** Informal unit of time defined as the amount of time it takes light to travel across the size of a proton, equal to about $3 \times 10^{-24}$ seconds.

**Macho** Massive Compact Halo Object. It is believed that there may be a significant number of Machos in the form of Jupiter type objects spread throughout the galaxies that account for the dark matter.

**muon** An elementary particle that is very much like an electron, only is more massive.

**naked singularity** A singularity is the center point where all of the matter of a **black hole** eventually falls. According to Einstein's **General Theory of Relativity**, and many recent observations, this singularity is cloaked by the **event horizon**, so we cannot see it. If the **event horizon** did not exist the singularity would be exposed, i.e., naked.

**neutron star** The remnant of a **supernova** explosion, a neutron star is solid neutrons.

**oxidation** In reference to iron, and more commonly known as rust, it is the combination of iron atoms with oxygen atoms.

**photon** The basic unit of light. The light reflecting off of this page consists of photons, as does all other forms of electromagnetic radiation. Photons are also the exchange particles of the electromagnetic force.

**pion** Quark-antiquark pair, useful to describe nuclear interactions.

**positron** The positron is an antielectron, which has the same mass as an electron but opposite charge.

**pulsars** **Neutron stars** that emit energy along their magnetic axis. The magnetic axis rotates, and when it points to the Earth we observe a pulse of radiation, and see nothing otherwise. Since the rotation rate is constant, the net effect is a periodic pulsation of energy.

**quantum mechanics** The theory that describes the behavior of particles and high energy effects on the small scale. More than just a theory, quantum mechanics shows that, at the atomic level, nature is not deterministic.

**quasar** Quasi stellar object, now believed to be galaxies with a supermassive **black hole** in their interior, which accounts for the enormous amount of radiation they emit.

**radiation pressure** The force of light, or any electromagnetic radiation. NASA has proposed solar sailboats, that use large aluminum "sails" that use the radiation pressure of the solar light to navigate through the solar system.

**radioactivity** The spontaneous emission of energy from matter. Usual examples are the emission of electrons, **photons**, or alpha particles from nuclei.

**red giants** Stars like our Sun that are in their death throes. As the last layers undergo **fusion**, the star increases its diameter by a factor of 100, making it a red giant.

**redshift** The increase in wavelength due to motion away from the observer. Distant galaxies are moving away from us, and therefore their light is redshifted.

**relativistic length contraction** The length of an object is not an absolute quantity, but depends on the velocity between it and the observer. A person traveling by a meter stick at a speed of $0.99c$ ($c$ is the speed of light) would measure it to be 14 centimeters long.

**Riemann tensor** In geometry, this quantity signals **curved space**. If it is zero, space is flat (Euclidian geometry is valid) and if it is not zero, space is curved. In physics, this quantity signals the presence of matter. It is only zero if no matter is present (the further the matter is removed, the smaller the Riemann tensor is). This object is a cornerstone in Einstein's **General Theory of Relativity**, which is why our **black hole** was trying to use it.

**SNOB** Society of Natural OBjects. All members agreed that there was a limit to how much information they would give to to our interviewer. SNOB is the metaphor for the limit that quantum theory places on the amount of information that exists in nature.

**string theory** A quantum theory in which particles are taken to be small strings. Usually formulated in terms of **supersymmetry** and more properly called superstring theory, this theory may automatically include a quantized theory of gravity. Some of its more intriguing predictions are that we live in a ten- or eleven-dimensional universe, and that every particle we have observed has a (as yet to be detected) superpartner.

**superconducting magnet** Certain elements and alloys, when cooled to a very low temperature, exhibit no resistance to electricity — a state called superconductivity. Superconducting magnets take advantage of the large currents, which create a magnetic field, that can be obtained in superconductors.

**supernova** The explosion of a star that was at least three or four times more massive than the Sun. The explosion occurs after the collapse of the star, which results from the termination of all **fusion** processes. The resulting explosion not only forms some of the heavier elements, but ejects them, along with other elements, across the universe. That is origin of the atoms that make this book you are reading.

**supersymmetry** A symmetry between bosons (spin one particles) and fermions (spin one half particles). A supersymmetric theory allows for the process whereby bosons can change into fermions and vice versa.

**tidal force** The gravitational force that tends to pull objects apart. The Moon's tidal force on Earth creates the tides, while tidal forces produced by solar mass **black holes** rip apart any object that gets too close. The origin of the tidal force arises from the difference in the strength of the gravitational field with distance: the side of the Earth closer to the Moon feels a stronger gravitational force than the side further away because the gravitational field of the moon, as for any object, falls off as the square of the distance.

**twenty-one-centimeter radiation** The radiation that hydrogen emits due to the spin flips of the electron.

**Wimp** Weakly Interactive Massive Particle, described in more detail by the neutralino.

**worm hole** Also known as the Einstein Rosen bridge, the worm hole is the curved region of spacetime that connects what appears to be two distinct **black holes**.

**X-ray burster** Objects that, from time to time, emit large bursts of x-ray radiation. It is now believed that this results from matter that builds up on a **neutron star**, and when it gets hot enough, undergoes **fusion**.

**x-ray radiation** A form of electromagnetic radiation (light is another form of electromagnetic radiation) which is characterized by a wavelength about one ten millionth of a millimeter. As gases rush into a **black hole**, their high speeds and collisions heat the gas which then emits x-rays, a signature of some **black holes**, just before they pass the **event horizon**.

**Z and W particles** Particles exert forces on each other by creating and absorbing exchange particles. The weak nuclear force is mediated by the Z and W particles. For example, when a neutrino interacts with an electron, a Z or W particle is created by the electron and absorbed by the neutrino, and vice versa. These creation and absorption processes cannot conserve both energy and momentum, so the exchange particles are called virtual particles.

**zero point energy** The lowest energy of a field. For example, the zero point energy of the electromagnetic field turns out to be infinite, and is usually throw away. Nevertheless, the zero point energy can be used to describe observable effects, but such effects are small.